CONCISE THERMODYNAMICS
Principles and Applications in Physical Science & Engineering

To
Faith, Jonathan and Bryony
who make it all worthwhile

Dr Samuel Johnson on lectures and books:

Talking of education, "People have now a-days" (said he,) "got a strange opinion that every thing should be taught by lectures. Now, I cannot see that lectures can do so much good as reading the books from which the lectures are taken. I know nothing that can be best taught by lectures, except where experiments are to be shewn. You may teach chymestry by lectures. — You might teach making of shoes by lectures!"

James Boswell: *Life of Samuel Johnson, 1766*

BENJAMIN THOMPSON, COUNT RUMFORD
(America)   1753-1814

NICHOLAS LEONARD SADI CARNOT
(France)   1792-1832

JAMES PRESCOT JOULES
(Germany)   1818-1889

RUDOLPH CLAUSIUS
(Germany)   1822-1888

WILLIAM THOMSON, LORD KELVIN
(Germany)   1822-1888

A Pantheon of great men of the past whose discoveries created thermodynamics as a
pervasive and interdisciplinary science in its own right

CONCISE THERMODYNAMICS
Principles and Applications in Physical Science and Engineering

JEREMY DUNNING-DAVIES
Department of Applied Mathematics
The University of Hull, Yorkshire

Albion Publishing
Chichester

First published in 1996 by
ALBION PUBLISHING LIMITED
International Publishers, Coll House, Westergate,
Chichester, West Sussex, PO20 6QL England

British Library Cataloguing in Publication Data
A catalogue record of this book is available from the British Library

ISBN 1-898563-15-2

Printed in Great Britain by Hartnolls, Bodmin, Cornwall

Contents

Preface

As its title indicates, this book is intended to be an introduction to the basic principles of thermodynamics. The material of the first eight chapters formed a one-term course given to final year undergraduates in applied mathematics here at Hull University. The enthusiasm and success of these students over a number of years provided a *raison d'être* for writing the book. The addition of chapters on phase transitions and questions of thermodynamic equilibrium and stability seem the natural way of extending the mentioned course to fit into the new semester system. The remaining chapters are concerned with research topics which have fascinated me recently. These could prove a source of material for a short postgraduate course, or may simply be of interest to research workers, particularly in physics and astrophysics.

Thermodynamics is the branch of science concerned with the ways in which the properties of matter and of systems change with alterations in temperature. It is a remarkable subject since it may be studied on both the microscopic and macroscopic levels; it applies to matter in all manner of extreme physical conditions; and yet, when examined in detail, is found to depend on "Laws" which are really only "facts of experience". Again, it is a topic which may be appreciated by people drawn from a wide variety of backgrounds: it is of importance to the physicist and astrophysicist, as well as the chemist; it is of tremendous importance to much that interests the engineer. At the same time, everyone , both young and old, meets examples of thermodynamics each day in the normal course of events. Thermodynamics is concerned with heat. Notions of "hot" and "cold", of one body being warmer than another, and the idea of the "flow of heat" are all central to the subject and, in science, **all** retain the meanings they have in our everyday lives. Initially, curiously enough, it is probably this latter point which is most difficult for many to accept but that is the absolute truth, thermodynamics **is** concerned with notions and concepts which are, in a non-scientific way, familiar to everyone. If this seemingly trivial point is borne in mind always, academic study of thermodynamics takes on a whole new perspective and is **not** a difficult subject to understand and appreciate.

The major part of this book is devoted to an examination of the basic laws and principles of thermodynamics. Topics such as phase transitions and questions of thermodynamic equilibrium and stability are also discussed, before attention is shifted to one or two topics which have concerned me personally over recent years. The latter involve looking at the mathematical property of *concavity* of the entropy; where the *entropy* is a function quite fundamental to the subject and is introduced via the Second Law of Thermodynamics - but more of this later! The text is completed with a critical examination of black hole thermodynamics, and a suggestion for a possible alternative model for a black hole is advanced. This final topic may be regarded by some as controversial since the view expressed disagrees with Hawking, - especially as far as the expression, associated with his name, for the entropy of a black hole is concerned. However, the arguments put forward are easily understandable to anyone who has followed the earlier chapters, and grown to accept the almost all-embracing power of the laws of thermodynamics. These two final topics take the book outside the realm of a purely undergraduate text, yet all the material included is accessible to undergraduates while possibly being of some interest to people at all levels.

I started to learn about thermodynamics as a postgraduate student of Peter Landsberg, now of Southampton University, and have continued to benefit from his knowledge and experience. I must thank him also for the photographs of some of the famous founding fathers of thermodynamics, which appear in this text. More recently, I have benefitted tremendously from the friendship of, and collaboration with, Bernard Lavenda of the University of Camerino in Italy, and from the friendship and encouragement of George Cole here in Hull University. Again, I am extremely grateful to Steve Hunter of Sheffield University for perusing a draft of this book with such care; he certainly helped eliminate several errors but any remaining must be my responsibility entirely. Also, I feel that, technically, I could not have produced this text without the help of Gordon McKinnon in resolving various word processing problems.

I gratefully acknowledge the financial support of the EU Third Framework 'Human Capital and Mobility Programme' (contract number CHRX - CT92-0007) and of the University of Hull *via* the award of a research support grant.

The entire project only started because of the publishing enthusiasm of Mr. Ellis Horwood, to whom I owe thanks for initiating the book and guiding me through all the pitfalls so carefully. Finally and above all, my grateful thanks to my wife, Faith, and children, Jonathan and Bryony. They have supported me at all times and have been 'forced' to listen to readings on various aspects of thermodynamics and black holes for quite some time.

Jeremy Dunning-Davies,
February 1996, University of Hull.

ABOUT OUR AUTHOR

Jeremy Dunning-Davies was born in Wales and attended Barry Boys' Grammar School, Glamorganshire (1951-1959). He went on to study mathematics in Liverpool University where he gained his B.Sc. in 1962, and a Certification of Education in 1963. He then undertook postgraduate study at the University College of Wales in Cardiff where he became a research student of Professor Peter Landsberg. It is not surprising that he developed a deep and lasting interest in the science of thermodynamics under his famous mentor. He was awarded his PhD for his thesis on "The ideal relativistic quantum gas" in 1966, and that same year entered the teaching profession with his appointment as Assistant Lecturer in the University of Hull, with successive promotions to full Lecturer in 1968 and Senior Lecturer in 1981

Meanwhile his investigative work deepened as he focused attention on the ideal Bose-Einstein gas, awakened by the comological implications of a massive primiordial gas and the growing problems of quarks and gas confinement. Together with Peter Landsberg he found that instead of using Boltzmann's principles to relate the entropy to "thermodynamic" probability, the entropy determines the form of the error law. Where Stirling's approximation is applicable, the probability distribution is a function of the difference between the entropy and its maximum value at equilibrium for which the average and most probable values coincide. This approach was used to establish that there are no intermediate statistics beetween the well-known Fermi and Bose statistics which are governed by the binomial and negative binomial distributions respectively. With Peter Landsberg again, followed a study of the Bose gases, and on statistical mechanics of quarks and hadrons, Dr Dunning-Davies also worked with his research student Dr. David Pollard on the gravitational action on a column of gas. He next became involved with the Caratheodory analytical approach and the connection between the First, Second and Third Laws of thermodynamics, and with negative absolute temperatures.

More recent research has been with Professor Bernard Lavendra of The University of Camerino in Italy on a probabilistic approach to thermodynamics, showing that physical statistics may have arisen from error laws belonging to exponential families of distributions. His impressive output of 75 published research papers in many fields of science, mostly on thermodynamics, statistical thermodynamics, atomic physics and electronic engineering, have deservedly confirmed the world status of this distinuished scientist.

1

Introduction

Thermodynamics is the branch of science which considers how changes of temperature affect the various properties of matter and of systems. The subject may be viewed on a microscopic level, in which case the interactions of atoms and molecules are studied as the temperature alters. For such a study, a specific model for the phenomenon under consideration is required. However, the truly unique status of classical thermodynamics becomes apparent when the subject is viewed macroscopically. In this case, only the behaviour of matter and radiation in bulk is considered: any internal structure they may possess is ignored. Hence, classical thermodynamics is concerned solely with relations between macroscopic observable quantities.

In many physical problems, details of the correct microscopic physics may not be known, but the thermodynamic approach may still provide answers about the macroscopic behaviour of the system - answers which are independent of the unknown detailed physics. In fact, thermodynamic arguments have absolute validity independent of the actual model used to explain any particular phenomenon.

It is remarkable to realise that these far-reaching statements may be made on the basis of the four laws of thermodynamics - the first two of which are arguably more important than the remainder. These laws themselves are remarkable also in that, in reality, they are no more than reasonable hypotheses formulated as a result of practical experience. Nevertheless, they prove to be of immense power, having been applied successfully to matter in extreme physical conditions; such as matter in bulk at nuclear densities inside neutron stars and in the early stages of the hot big bang model of the Universe, as well as at very low temperatures in laboratory experiments. However, there is no way in which the laws of thermodynamics may be proved - they are simply expressions of common experience of the thermal properties of matter and radiation.

From the outset, it is important for the student new to this field to realise that by 'common experience' is meant simply what it states: the experiences which form the basis of thermodynamics are ones with which, in one way or another, everyone is familiar. Words such as 'hot' and 'cold' retain their everyday meanings ; the idea of one body being 'hotter' than another is familiar to anyone who has inadvertently touched a heated towel rail; the idea of heat flowing from a body to a colder one is, again, a concept familiar to all who have wished to become warm after venturing out on a cold winter's day and have gained comfort from sitting close to a roaring fire. These simple experiences are at the heart of the subject and, when faced with a problem of understanding in thermodynamics, students would do well to remember these simple everyday occurrences with which everyone is familiar. Concepts such as temperature and pressure (where pressure, as usual, is simply the force acting on unit area) also retain their everyday meanings and students should bear this in mind when meeting these in future thermodynamic discussions.

However, while thermodynamics *is* rooted in experiences which are familiar to all, some more advanced aspects of the subject place it among the most abstract branches of physics. Although rarely emphasized, this is an important point to note, since the main reason for it is that the basic theory contains results which, within broad limits, are independent of any particular system. This leads to the surprisingly wide range of applications for thermodynamic results: for example, as well as being of obvious use in physics and chemistry, parts of the theory find application in biology, information theory, communications, and even the study of language.

Anticipating what is to follow, it might be noted that the First Law of Thermodynamics concerns conservation of energy and may be stated as

Energy is conserved when heat is taken into account.

The Second Law gives information concerning the way in which systems evolve. There are several statements of this particular law but that due to Clausius is

No process is possible whose only result is the transfer of heat from a colder to a hotter body.

These days it is probably the Second Law which causes more problems of understanding but, historically, it was the First Law which easily proved the more difficult to establish - possibly due the difficulty of understanding the precise nature of heat. In the 18th Century, heat was regarded as some sort of massless fluid, called **caloric**. It was thought that when one body is at a higher temperature than another and both are brought into thermal contact, caloric would flow from the hotter to the colder body until they came to equilibrium at the same temperature. However, this theory had problems. For example, when a warm body is brought into contact with ice, caloric will flow from the warm body to the ice; but, although ice is converted into water, the temperature of the ice-water mixture remains unaltered. Also in the 18th Century, an alternative view developed according to which heat is associated with the motions or vibrations of the microscopic particles which make up matter. This theory - the so-called **kinetic theory** - *associated heat with the kinetic energy of the motions of the microscopic constituents of matter.*

The two theories came into conflict at the end of the century when the experiments of Count Rumford weighed heavily against the caloric theory. Rumford was an American who moved to Britain during the American War of Independence.He was made a count of the Holy Roman Empire for service in Bavaria in 1791, and was also something of a soldier of fortune. In 1798, by attempting to bore cannon with a blunt drill, he showed that heat could be produced by friction. In this experiment there is no obvious source of caloric, which apparently may be produced in indefinite amounts.

A major breakthrough came in 1822 with the publication of Fourier's *Analytical Theory of Heat.* In this book, Fourier evaluated the mathematical theory of heat transfer in the form of differential equations which did not require the construction of a specific model of the physical nature of heat. His methods gave mathematical expression to the *effects* of heat without enquiring into its causes. Also, and most importantly for future developments, this process of assigning mathematical substance to physical effects profoundly influenced the future generation of Scottish physicists - in particular William Thomson, who was later to become Lord Kelvin.

By the 1820's, the relation between kinetic energy and work done had been clarified. Then, in the 1840's, a number of scientists independently came to the correct conclusion concerning the interconnectibility of heat and work. In 1842, Mayer proposed that heat and work are interchangeable and proceeded to derive a value for the mechanical equivalent of heat from the adiabatic expansion of gases. However, the most important contributions were to come from another Scot, James Prescott Joule, who carried out a superb series of experiments which was crucial to the foundation of the law of conservation of energy.

Joule came from a family which had become wealthy through the brewery it owned and ran. He himself was really an amateur scientist and performed his experiments in laboratories installed and equipped at his own expense in his home and at the brewery. His genius was as a truly meticulous experimenter and the reason for singling him out for special mention is that he, above all others, gave thermodynamics a sound experimental basis. He established experimentally that the different forms of energy - heat, mechanical energy, and electricity - are equivalent and may be converted one into another, confirming the principle of conservation of energy. An important aspect of his work was the ability, even in the mid 19th century, to measure very small temperature changes very accurately. Indeed, by taking the utmost care to estimate all heat losses, he derived a value for the mechanical equivalent of heat of 4.13 Joules/calorie. This should compared with the modern accepted value of 4.187 Joules/calorie.

The early results of Joule's experiments greatly excited Thomson who appreciated their significance immediately and used them as the foundation for the subject now called thermodynamics. The basic results became widely known and, by 1850, Helmholtz and Clausius had formulated what is known now as the **law of conservation of energy** or the **First Law of Thermodynamics**.

2

The Zeroth Law

As has been mentioned already, everyone is familiar with such elementary notions as 'A is warmer than B', 'B may gain heat from A', and the qualitative notion of the 'flow of heat'. Also, everyone knows that, when the flow of heat between two systems has ceased, those systems are said to be in thermal equilibrium. The above might all be termed *'facts of experience'* and, in many arguments which lead to the formulation of the laws of thermal physics, such 'facts of experience' play an important role.

Now, if the influence of, for example, electric or magnetic fields is absent, it is a fact of experience that the properties of a stationary fluid are determined completely by just two properties - the pressure p and volume of the containing vessel V. A system defined by only two properties is termed a two-coordinate system. Such systems occur widely but the formalism may be generalised easily to cope with multi-coordinate systems.

Consider two isolated systems consisting of fluids with coordinates p_1, V_1 and p_2, V_2. If brought into thermal contact and left for a long time, the properties of these two systems will change so that a state of thermal equilibrium is achieved. This means that all components making up the system are allowed to interact thermally until, after a long time, no further changes are observed in the bulk properties of the system. Generally, heat will be exchanged and work done in attaining this final situation. Eventually, the two come to thermal equilibrium such that their thermodynamic coordinates assume the values p_1', V_1' and p_2', V_2'. However, it is a fact of experience that the four coordinates cannot be totally independent if the two systems are in thermal equilibrium. Hence, there must be some relation linking these four quantities:

$$F(p_1, V_1, p_2, V_2) = 0.$$

This equation enables one quantity to be found in terms of the other three.

So far, use of the word 'temperature' has been avoided but a suitable definition may be found by using another 'fact of experience' which is so central to the subject that it appears as a law of thermodynamics - the **Zeroth Law**. This states that

If two systems, 1 and 2, are separately in thermal equilibrium with a third 3, then they must be in thermal equilibrium with one another.

Systems 1 and 3 being in thermal equilibrium means a relation of the form

$$F(p_1, V_1, p_3, V_3) = 0$$

holds. This may be expressed alternatively as

$$p_3 = f(p_1, V_1, V_3).$$

Similarly, systems 2 and 3 being in thermal equilibrium implies $p_3 = g(p_2, V_2, V_3)$. Hence,

$$f(p_1, V_1, V_3) = g(p_2, V_2, V_3). \tag{2.1}$$

However, according to the Zeroth Law, systems 1 and 2 must be in thermal equilibrium also and so, a relation of the form

$$H(p_1, V_1, p_2, V_2) = 0$$

must hold.

This implies that (2.1) must be of a form such that the dependence on V_3 on either side cancels; that is, for example,

$$f(p_1, V_1, V_3) = \phi_3(p_1, V_1)\xi(V_3) + \eta(V_3)$$

and

$$g(p_2, V_2, V_3) = \phi_2(p_2, V_2)\xi(V_3) + \eta(V_3).$$

Hence, if the dependence on V_3 is cancelled out, it is seen that, in thermal equilibrium

$$\phi_1(p_1, V_1) = \phi_2(p_2, V_2) = \phi_3(p_3, V_3) = t = \text{constant}.$$

This is the logical consequence of the Zeroth Law - there exists a function of p and V which may well vary in form from one system to another but which takes a constant value for *all* systems in thermal equilibrium with one another. Different equilibrium states will be characterised by different constants. This constant , which characterises the equilibrium, is termed a **function of state** - that is, it is a quantity which assumes a definite value for a particular equilibrium state - and is called the **empirical temperature** t. (Here it might be noted that the word *empirical* means based on observation and experiment, not on theory.)

An equation of state relating p and V to this empirical temperature emerges also

$$\phi(p, V) = t.$$

All the combinations of p and V which correspond to a fixed value of the empirical temperature t may be found from experiment. Of these three quantities, any two are sufficient to define the equilibrium state completely. Lines of constant t plotted on a pressure - volume, (p-V), diagram are called **isotherms**.

At this stage, the empirical temperature looks nothing like what is normally called temperature. To place everything on a firm experimental foundation, a thermometric scale must be chosen. Once that is fixed for one system, it will be fixed for all others since all systems must have the same value of the empirical temperature when they are in thermal equilibrium.

Now consider a device which, in attaining thermal equilibrium with a physical system, disturbs that system to a negligible extent. Also, suppose that all its thermodynamic variables except one are constrained to practically fixed values. When this device is allowed to reach thermal equilibrium with a system, ideally its variable characteristic is a strictly increasing or strictly decreasing function of the empirical temperature. Hence, the device may be calibrated to read temperature directly and is a *thermometer*. If a gas is used as the medium, the thermometer may be of the constant volume or constant pressure variety.

Note that the notions of 'hot' and 'cold' may be associated with high and low temperatures respectively - but the reverse may be true equally well. As yet, it is not required to restrict the choice of an empirical temperature and so, both possibilities must be allowed.

Exercises A

Familiarity with partial differentiation is crucial for many thermodynamic manipulations. Therefore, this first set of examples is intended to help with revision of the techniques of partial differentiation which are so important in the present context. The main basic results required are derived in the appendix.

(1)   Given $u = x^2 + 2x - 1$ and $x = t^2 - 1$, find du/dt by
       (a)   substituting t for x,
       (b)   using the chain rule.

(2)   Given $u(x,y) = (x + 1)^2 - 3xy^2 + 4y$, find
       (a) $u(2,-1)$, (b) $u(1/x, x/y)$.

(3)   Find all the first partial derivatives of
       (a) $u(x,y) = \tan(x/y)$
       (b) $f(r,\theta) = r^2\sin^2\theta + r^3$
       (c) $u(r,s,t) = r^3 + s^2t + (t - 1)(r - 3)$
       (d) $f(p,q) = \exp(p^2\log q)$.

(4)   The relation between the pressure p, temperature t and volume V of a certain amount of hydrogen gas may be expressed over a limited range by the equation
$$p(V - B) = At$$
where B is independent of p but is a function of t ; A is a constant. Find $\left(\partial V / \partial t\right)_p$ and $\left(\partial V / \partial p\right)_t$ and express dV as a function of t and p.

(5)   If $z = f(x,y)$, prove the important result
$$\left(\frac{\partial x}{\partial y}\right)_z \left(\frac{\partial y}{\partial z}\right)_x \left(\frac{\partial z}{\partial x}\right)_y = -1.$$
This result should be known; it will prove to be of use on innumerable occasions.
[Here the symbol outside each bracket indicates that that variable is the one to be held constant in the differentiation. This particular element of notation is peculiar to physics and to thermodynamics in particular. It does have the merit of making it quite clear which variable is being held constant.]

(6)   An increment
$$d'Z = L(x,y)dx + M(x,y)dy$$
is an exact differential if there exists a function $Z(x,y)$ such that
$$L(x,y) = \left(\partial Z / \partial x\right)_y ; \quad M(x,y) = \left(\partial Z / \partial y\right)_x .$$

A necessary and sufficient condition for $d'Z$ to be exact is

$$(\partial L / \partial y)_x = (\partial M / \partial x)_y \ .$$

The volume of a figure is given by $V = ar^b h^{3-b}$, where a and b are constants. Find $(\partial V / \partial h)_r$ and $(\partial V / \partial r)_h$. Give an expression for dV in terms of dr and dh, and verify that dV is exact.

(7)   Verify that $\dfrac{\partial^2 f}{\partial x \partial y} = \dfrac{\partial^2 f}{\partial y \partial x}$ for

(a) $f(x,y) = \sin^2(x+y) + x^2 \cos y$
(b) $f(x,y,z) = x^3 y^2 z$

(8)   If the variables x, y and η, ξ are related by

$$x = x(\eta, \xi) \ ; \ \ y = y(\eta, \xi)$$

the Jacobian is defined to be

$$\frac{\partial(x,y)}{\partial(\eta,\xi)} = \begin{vmatrix} \left(\dfrac{\partial x}{\partial \eta}\right)_\xi & \left(\dfrac{\partial y}{\partial \eta}\right)_\xi \\ \left(\dfrac{\partial x}{\partial \xi}\right)_\eta & \left(\dfrac{\partial y}{\partial \xi}\right)_\eta \end{vmatrix} .$$

It follows that

$$\frac{\partial(x,y)}{\partial(x,y)} = -\frac{\partial(y,x)}{\partial(x,y)} = 1$$

and, if u is a function of x and y also,

$$\frac{\partial(u,y)}{\partial(x,y)} = \left(\frac{\partial u}{\partial x}\right)_y .$$

Also note that

$$\frac{\partial(u,y)}{\partial(x,y)} = \frac{\partial(u,y)}{\partial(x,z)} \frac{\partial(x,z)}{\partial(x,y)} .$$

Using the above results, show that, if E is a function of t, N, and μ, and N itself depends on t and μ,

$$\left(\frac{\partial E}{\partial t}\right)_\mu = \left(\frac{\partial E}{\partial t}\right)_N + \left(\frac{\partial E}{\partial N}\right)_t \left(\frac{\partial N}{\partial t}\right)_\mu .$$

The basic results introduced in questions (6) and (8) should be noted for future use; both can be extremely useful.

(9)   Show that, if $z = f(x^n y)$ where $n \neq 0$, then

$$x \frac{\partial z}{\partial x} = n y \frac{\partial z}{\partial y}.$$

3

The First Law.

A simple straightforward statement of this law has been given already; that is,

Energy is conserved when heat is taken into account.

However, to give real meaning to this, heat, energy and work must be defined clearly. The first two have been met already but what about work? Work has a different meaning in science from its everyday one. In fact, work is done when a force moves, so that, if someone lifts a pile of books, work is done since the force acting on the books actually moves; but, if someone simply holds a pile of books, no work is done because, although an upward force is exerted, no motion results and the force does not move. The amount of work done in this simple example will depend on both the number of books to be lifted and the height through which they are lifted. Hence, work is measured by the product of the force acting and the distance moved in the direction of the force. In mathematical terms, therefore, if a force F acts on a body which is moving along the path $r = r(t)$, the work done in a small displacement dr is defined to be $F.dr$, and so the total work done in moving the body from $r = r_1$ to $r = r_2$ is given by

$$W = \int_{r_1}^{r_2} F.dr$$

and when work is done *on* a body, the energy of that body is increased. Now consider a few examples:

The work done in compressing a fluid.

When work is done on a fluid, the work done is regarded *by convention* as being positive. Then, since the volume decreases when mechanical work is done to compress the fluid, the work done is

$$W = -\int p dV$$

where p is the pressure of the fluid and V is the volume. Note that, if the fluid does work by expanding, the work done on the surroundings is positive and that done on the fluid is negative. Hence, the sign of the volume increment is important.

Work done on a charge q by an electric field.

In this case, if the electric field is E and the charge moves through a distance dr, the work done is given by

$$dW = qE.dr$$

Work done by a magnetic field.

In this case, if B is the magnetic field and m the magnetic dipole moment, an increment of work done by the magnetic field is given by

$$dW = B.dm$$

These are simply examples quoted to illustrate the form taken by the expression for the work done in a variety of cases. It may be noted that, in each case, the work done is the product of a generalised force X and a generalised displacement dx, so that

$$dX = X.dx.$$

Now consider an isolated system in which there is no thermal interaction with the surroundings. It is a 'fact of experience' that, if work is done on the system in some way, the system attains a new equilibrium state and it does not matter how the work which achieves this is done: for example, a gas may be compressed, or stirred, or have an electric current passed through it. It was one of Joule's great contributions to thermodynamics to demonstrate experimentally that this is the case. The result is that energy is given to the system during the process and, since no thermal interaction is involved, the process is said to be *adiabatic*. It follows that, if a system is caused to change from some initial state to a final state by adiabatic means, the work done is the same no matter how it is done.. Hence, there must exist a function of the coordinates of the system whose value in the final state minus its value in the initial state equals the work done in going from one state to the other. This function of state is called the *internal energy* and is denoted by U. For the isolated system

$$W_a = U_2 - U_1 \qquad\qquad (3.1)$$

where U_2 and U_1 are the final and initial values respectively of the internal energy and W_a is the work done in this adiabatic process; the suffix a indicating that the process is adiabatic.

Suppose now that the system is not isolated as above but that thermal interaction between the system and its surroundings is allowed. In this case, the system may be taken from the state with internal energy U_1 to that with internal energy U_2 by a process which is not necessarily adiabatic. Such a process may be achieved by performing work which may be mechanical - for example, the use of a stirrer , non-mechanical - for example, the use of a heating element, or a combination of the two. Let W_{na} denote the mechanical work done on the system in a process which is not necessarily adiabatic; the suffices na indicating that the process is not necessarily adiabatic. Then

$$W_a - W_{na} = Q \tag{3.2}$$

and for all such processes

$$Q + W_{na} = U_2 - U_1 . \tag{3.3}$$

Here Q is zero for adiabatic processes only. In a non-adiabatic process, Q may be thought of as making up the deficit of mechanical work by heat. Hence, the amount of heat is defined in terms of mechanical work only. The convention adopted is that positive values of Q will mean heat supplied *to* the system. Also, it should be noted that, although it has not been stated explicitly, attention has been confined to *closed* systems; that is, systems which do not transmit mass to, or receive mass from, the surroundings.

On occasions, it proves useful to consider equation (3.3) written in differential form:

$$d'Q = dU - d'W . \tag{3.4}$$

Here dU is the difference in internal energy between the two states and is a differential of a function of state. However, $d'Q$ and $d'W$ are not differentials of functions of state - the system may be taken from state 1 to state 2 by adding different amounts of heat and mechanical work. The dashes on $d'Q$ and $d'W$ are to draw attention to this and to the fact that neither turns out to be an exact differential.

Consider now the idea of a so-called perpetual motion machine of the first kind. Such a machine would be a system which interacts with its surroundings only mechanically and does a positive amount of work on the surroundings during a closed transition; where a closed transition is one for which the initial and final states are the same, that is, $U_2 = U_1$. However, by equation (3.3), this is seen to be impossible to achieve since the simultaneous conditions $U_2 = U_1$ and $Q = 0$ are incompatible with $W_{na} > 0$. Hence, the First Law declares the impossibility of constructing perpetual motion machines of the first kind. The validity of the First Law is supported by the fact that, so far, all attempts to construct such machines have failed.

Some Applications of the First Law

Specific heat capacities

As mentioned already, the internal energy, U, is a function of state and, for a two-coordinate system such as a perfect, or ideal, gas, properties may be described in terms of two other functions of state. Suppose the volume, V, and empirical temperature, t, are taken as these two independent variables so that

$$U = U(V,t)$$

and

$$dU = \left(\frac{\partial U}{\partial V}\right)_t dV + \left(\frac{\partial U}{\partial t}\right)_V dt .$$

Then, since , as was seen earlier, $d'W = -pdV$,

$$d'Q = \left(\frac{\partial U}{\partial t}\right)_V dt + \left\{\left(\frac{\partial U}{\partial V}\right)_t + p\right\}dV .$$

The heat capacity at constant volume is given by

$$C_V = \left(\frac{\partial' Q}{\partial t}\right)_V = \left(\frac{\partial U}{\partial t}\right)_V$$

and the heat capacity at constant pressure is given by

$$C_p = \left(\frac{\partial' Q}{\partial t}\right)_p = \left(\frac{\partial U}{\partial t}\right)_V + \left\{\left(\frac{\partial U}{\partial V}\right)_t + p\right\}\left(\frac{\partial U}{\partial V}\right)_p .$$

These expressions for the two heat capacities show how much the temperature rises for a given input of heat under the conditions of constant volume and constant pressure respectively. It will be found that both these quantities occur frequently in thermodynamic manipulations and calculations. However, the two expressions given do not refer to any particular volume or mass. Often it proves convenient to use the so-called *specific heat capacities* or *specific heats*, which are heat capacities per unit mass; that is

$$c_V = C_V/m \quad \text{and} \quad c_p = C_p/m .$$

It follows immediately from the above expressions for the two heat capacities that

$$C_p - C_V = \left\{\left(\frac{\partial U}{\partial V}\right)_t + p\right\}\left(\frac{\partial V}{\partial t}\right)_p .$$

It might be noted at this stage that an ideal, or perfect, gas is one satisfying the equation

$$pV = Rt$$

where R is the gas constant for one mole of gas, and

$$\left(\partial U / \partial V\right)_t = 0 ,$$

this second equation in the definition of an ideal gas stemming from an experiment by Joule in which a compressed gas was released into a vacuum. Hence, for an ideal gas

$$C_p - C_V = p\left(\partial V / \partial t\right)_p = R .$$

It might be noted also that a *mole* is defined as the amount of substance that contains as many particles (whether atoms, ions or molecules) as exactly 13gms. of carbon-12. The number of atoms in 12gms. of carbon-12 is found by experiment to be 6.02×10^{23};-a number known as Avogadro's constant, after the nineteenth century Italian chemist Amadeo Avogadro.

The enthalpy and the Joule-Kelvin expansion

In the following discussion, p and t are taken to be the independent variables so that, here, U is of the form

$$U = U(p,t)$$

and

$$dU = \left(\frac{\partial U}{\partial p}\right)_t dp + \left(\frac{\partial U}{\partial t}\right)_p dt \; .$$

Then, as before

$$d'Q = dU + pdV$$

$$= \left(\frac{\partial U}{\partial p}\right)_t dp + \left(\frac{\partial U}{\partial t}\right)_p dt + pdV$$

and

$$\left(\frac{\partial' Q}{\partial t}\right)_p = \left(\frac{\partial U}{\partial t}\right)_p + p\left(\frac{\partial V}{\partial t}\right)_p$$

$$= \left(\frac{\partial (U + pV)}{\partial t}\right)_p \; .$$

$(U + pV)$ is composed entirely of functions of state and so, must itself be a function of state. It is known as the **enthalpy**, is denoted by H and

$$H = U + pV.$$

It follows immediately that

$$C_p = \left(\frac{\partial' Q}{\partial t}\right)_p = \left(\frac{\partial H}{\partial t}\right)_p \; .$$

The enthalpy often appears in flow processes and, in particular, occurs when discussing the Joule-Kelvin (*sometimes called the Joule-Thomson*) expansion process. In this process, gas at a relatively high pressure, say p_1, is allowed to stream through a throttling valve into a region of lower pressure, say p_2, in a continuous stream. The expansion through the valve is irreversible and, after emerging from the valve, the gas is in a state of turbulent flow. It soon comes to an equilibrium state at a lower pressure and is found to have changed its temperature. To make the approach to equilibrium as rapid as possible, the valve is replaced by a porous plug, which removes all irregular currents from the gas before it emerges. It is necessary to obtain a steady flow and to measure the pressure and temperature differences on the two sides of the plug. If the pressure and temperature changes are Δp and Δt respectively, the Joule- Kelvin coefficient is defined to be $\Delta t / \Delta p$.

The experimental situation is illustated in Figure (3.1):

direction of flow

$\rightarrow$

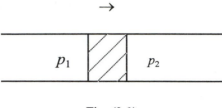

Fig. (3.1)

Suppose the volume of gas pushed into the pipe at pressure p_1 is V_1. Since p_1 is constant throughout this pipe, the work done on this sample of gas is

$$\int p_1 dV_1 = p_1 V_1.$$

After passing through the plug, the same mass of gas has volume V_2 and , since its new pressure is p_2, does work $p_2 V_2$ in passing out of the pipe. Thus, the external work done by the gas during the entire process is

$$p_2 V_2 - p_1 V_1.$$

It is assumed that no heat is absorbed, so that , if U_1 is the internal energy when the gas enters the pipe and U_2 when it leaves, the First Law gives

$$U_2 - U_1 = -(p_2 V_2 - p_1 V_1)$$

that is

$$U_2 + p_2 V_2 = U_1 + p_1 V_1$$

or

$$H_2 = H_1.$$

Therefore, the change takes place at constant enthalpy and the Joule-Kelvin coefficient is $(\partial t / \partial p)_H$.

The above comes close to deriving a more general flow equation in which account is taken of other contributions to the flow; for example, the kinetic energy and also the potential energy of the gas if it is in a gravitational field. Consider the flow through a black box and, in particular, the steady flow of a given mass of gas or fluid as it enters and leaves the black box. Conservation of energy gives

$$H_1 + \tfrac{1}{2} m v_1^2 + m\phi_1 = H_2 + \tfrac{1}{2} m v_2^2 + m\phi_2$$

that is

$$U_1 + p_1V_1 + \tfrac{1}{2}mv_1^2 + m\phi_1 \;=\; U_2 + p_2V_2 + \tfrac{1}{2}mv_2^2 + m\phi_2$$

that is

$$U + pV + \tfrac{1}{2}mv^2 + m\phi \;=\; \text{constant} \tag{3.5}$$

where v is the velocity and ϕ the gravitational potential.

This may be written

$$p/\rho + u + \tfrac{1}{2}v^2 + \phi \;=\; \text{constant}$$

where $\rho = m/V$ is the mass density and u the energy density. This is leading towards the equations of fluid flow. In fact, for an incompressible fluid, $U = $ constant and the above equation becomes

$$p/\rho + \tfrac{1}{2}v^2 + \phi \;=\; \text{constant,}$$

which is Bernoulli's equation.

It might be noted that these considerations of a more general flow equation (3.5) show that the Joule-Kelvin process must be assumed to proceed very slowly so that the kinetic energy term may be neglected.

Adiabatic expansion

In an adiabatic expansion, the volume of the gas changes with *no* thermal contact between the system and its surroundings. An important point to note about processes of this type is that they proceed very slowly so that the system passes through an infinite number of equilibrium states in passing from the initial state to the final state. Obviously these processes are ideal and cannot actually occur in practice. However, it is vital to realise that such processes can be approximated to very closely, so considering them is not too far removed from reality. Consider the expansion to be for an ideal gas for which, as seen earlier,

$$pV = Rt \quad \text{and} \quad C_p - C_v = R.$$

Then, the First Law gives

$$dU + pdV = d'Q = 0. \tag{3.6}$$

Also, since for an ideal gas

$$\left(\partial U / \partial V\right)_t = 0,$$

$$C_v = \left(\partial U / \partial t\right)_v \Rightarrow dU = C_v dt.$$

Hence, equation (3.6) leads to

$$0 \; = \; C_v dt \; + \; pdV \; = \; C_v dt \; + (Rt/V)dV$$

or

$$\left(\frac{C_v}{C_p - C_v} \right) \frac{dt}{t} = -\frac{dV}{V}$$

that is

$$\frac{1}{\gamma - 1} \frac{dt}{t} = -\frac{dV}{V} \tag{3.7}$$

where γ is the ratio of the constant pressure and constant volume heat capacities.

Integrating (3.7) gives

$$Vt^{1/(\gamma - 1)} = \text{constant}$$

or, more familiarly,

$$pV^{\gamma} = \text{constant,}$$

where, once again, $pV \; = \; Rt$ has been used.

Isothermal expansion

Consider the expansion or compression of the gas in a cylinder as illustrated in Figure (3.2).

Fig.(3.2)

In this case, heat exchange with the surroundings must occur so that the gas inside the cylinder remains at the same constant temperature - a situation demanded by the process being isothermal. Here work is done in pushing back the piston and this is compensated by a corresponding inflow of heat.

$$\text{Work done} \; = \; \int_{V_1}^{V_2} pdV = \int_{V_1}^{V_2} \frac{Rt}{V} dV = Rt \ln\left(V_2 / V_1 \right).$$

This is the amount of heat which must be supplied from the surroundings to maintain an isothermal expansion.

Types of expansion

It seems useful, at this point, to summarise briefly the properties of the three important types of expansion met earlier in this chapter:

Isothermal expansion $\Delta t = 0$. Heat must be supplied or removed to maintain $\Delta t = 0$

Adiabatic expansion $\Delta Q = 0$. No heat exchange occurs with the surroundings.

Joule-Kelvin expansion $\Delta H = 0$. When gas passes from one volume to another and the pressures in the two vessels are maintained at p_1 and p_2 during the transfer enthalpy is conserved if the gas is ideal.

In all these apparently different phenomena, the basic principle involved is the First Law of Thermodynamics; that is, all involve conservation of energy with account being taken of heat.

Exercises B

(1)   The van der Waals' equation of state is

$$\left(p + \frac{a}{V^2}\right)(V - b) = At$$

where a, b and A are constants.

In a p-V diagram, the stationary points lie on a curve. Find the equation of this curve and show that its maximum is given by

$$V_{cr} = 3b \; ; \quad p_{cr} = a/27b^2 \; ; \quad t_{cr} = 8a/27Ab.$$

Further show that the van der Waals' equation may. be written

$$\left(\pi + 3/\phi^2\right)\left(3\phi - 1\right) = 8\tau$$

where $\phi = V/V_{cr}, \pi = p/p_{cr}, \tau = t/t_{cr}$

(2)   In a general equation of state,

$$pV = A + Bp + Cp^2 + ...$$

is written sometimes, where $A,B,C,...$ are functions of t and are called the **first, second, third,... virial coefficients**.

If a,b in the van der Waals' equation of state are *small*, show that a gas obeying that equation has an approximate second virial coefficient

$$B = b - a/At.$$

(3)   The Boyle temperature, t_B, defined by $\left[\dfrac{\partial}{\partial p}(pV)\right]_{p=0} = 0$, is such that, near it, a general gas ($C,D,....$small) approximates Boyle's law $pV = A(t)$. Show that, for a van der Waals' gas,

$$t_B/t_{cr} = 3.375$$

In the following examples, only the First Law of Thermodynamics will be required.

(4)   Consider the equation

$$d(\log V) = \alpha_p \, dt - \kappa_t \, dp$$

(a)   how must α_p and κ_t be defined for this result to hold?

(b)   interpret α_p and κ_t physically in words.

(c)   show that

$$\left(\frac{\partial \alpha_p}{\partial p}\right)_t = -\left(\frac{\partial \kappa_t}{\partial t}\right)_p$$

(5)   By regarding $d'Q$ as a function of t,V; t,p; V,p leads to

$$d'Q = C_V dt + l_V dV$$
$$= C_p \, dt + l_p \, dp$$
$$= m_V \, dV + m_p \, dp$$

where $C_V = \left(\dfrac{\partial' Q}{\partial t}\right)_V, C_p = \left(\dfrac{\partial' Q}{\partial t}\right)_p$ are the constant volume and constant

pressure heat capacities respectively; $l_V = \left(\dfrac{\partial' Q}{\partial V}\right)_t, l_p = \left(\dfrac{\partial' Q}{\partial p}\right)_t$ are the latent

heats   of   volume   and   pressure   increase   respectively;   and

$m_V = \left(\dfrac{\partial' Q}{\partial V}\right)_p, m_p = \left(\dfrac{\partial' Q}{\partial p}\right)_V.$

Using the above equations, show that

(a) $$m_V = \frac{l_V C_p}{C_p - C_V}, m_p = -\frac{l_p C_V}{C_p - C_V},$$

(b) $$\frac{m_V}{l_V} + \frac{m_p}{l_p} = 1.$$

{Note that a discussion of latent heats will be included at the beginning of Chapter 10.}

(6)   The Grüneisen ratio, Γ, is defined by

$$\Gamma = \frac{\alpha_p V}{\kappa_t C_V} \; .$$

Show that

$$\Gamma = \frac{V}{C_V}\left(\frac{\partial p}{\partial t}\right)_V = \frac{V}{(\partial U/\partial p)_V} = \frac{V}{m_p} \; .$$

(7)   For a fluid obeying Boyle's law, $pV = t$, and undergoing quasistatic adiabatic changes at constant γ $(=C_p/C_V)$, show that

$$tV^{\gamma-1} = const., \quad \text{and} \quad t^{\gamma}p^{1-\gamma} = const.$$

(8)   The enthalpy, H, is given by

$$H = U + pV.$$

Show that

(a) $dH = C_p\,dt + (l_p + V)dp$, where $C_p = \left(\partial H/\partial t\right)_p$.

(b) $\left(\dfrac{\partial U}{\partial V}\right)_t = \left(\dfrac{\partial p}{\partial V}\right)_t\left[\left(\dfrac{\partial H}{\partial p}\right)_t - V\right] - p$

(c) $\left(\dfrac{\partial U}{\partial V}\right)_t = -\left(\dfrac{\partial p}{\partial V}\right)_t\left[\mu C_p + V\right] - p$

where $\mu = \left(\partial t/\partial p\right)_H$ is the **Joule Thomson coefficient**.

(9)   Show that, if $pV = At$, with A a constant, then

$$\left(\frac{\partial U}{\partial V}\right)_t = \frac{p\mu C_p}{V} \; .$$

{It might be noted that, for a normal gas at low pressure, $pV = At$ is a good approximation and μ may be measured. Thus, if $\mu = 0$ is found at low pressures, $\left(\partial U/\partial V\right)_t = 0$, which is Joule's Law, may be inferred.}

(10)   For a classical ideal gas, for which $pV = At$ and Joule's Law holds, show that

$$d'Q = \frac{(n-1)C_V - A}{(n-1)} dt = \left[A - (n-1)C_V \right]\frac{t}{V} dV \, ,$$

if the change is such that $pV^n = const.$

{It might be noted that changes for which $pV^n = const.$ are called **polytropic** changes.}

4

The Second Law

Historically, the origin of the Second Law of Thermodynamics is linked with the name of Sadi Carnot. He was born in 1796, the eldest son of Lazare Carnot who was best known for his political activities. Lazare Carnot was a member of the Directory after the French Revolution - having previously been, amongst other things, a member of the notorious Committee of Public Safety - and, later, during the Hundred Days in 1815, Napoleon's Minister of the Interior. However, throughout his political career, he managed to find time for intellectual pursuits. His big interest appears to have been mechanics and, although he did little original work, it is felt nowadays that his attempt to produce a general science of machines did influence his son.

Sadi Carnot himself was educated at the élite École Polytechnique and, after a period as a military engineer, devoted himself to research. His great work, with English title *Reflexions on the motive power of fire,* was published in 1824. In modern terminology, motive power is work and the book was concerned with the maximum efficiency of heat engines. By the 1820's, with the restoration of peace between Britain and France, it became apparent that the French lagged a long way behind the British in some technological areas and nowhere was the disparity worse than in power technology. At that time, this area had become particularly important because of the widespread use of steam engines - in Britain, such machines were used, for example, in the Cornish tin mines both for pumping out water and for hauling men and loads of ore to the surface. The work of such engineers as Watt, Trevithick and Woolf was well-known and must have helped provide some inspiration and incentive for Carnot.

Quite naturally, for the time, Carnot adopted the so-called caloric theory in his work. This theory has been mentioned already in Chapter 1 and basically regards heat as some sort of massless fluid. Carnot assumed caloric conserved in the cyclic operation of heat engines and postulated that the origin of the work done by a heat engine is the transfer of caloric from one body to a colder body; - this flow of caloric being regarded as analogous to the flow of fluid which, as in a waterwheel, produces work when falling down a potential gradient. Crucially, Carnot recognised that a heat engine works most efficiently if the transfer of heat occurs as part of a cyclic process and also, that the main factor in determining the amount of work which may be extracted from a heat engine is the temperature difference between the heat source and the sink into which the caloric flows. Both these points turn out to be independent of the actual model of the heat flow process. Finally, he devised a cycle of operations - now known as the **Carnot cycle** - as an idealisation of the behaviour of any heat engine.

From these essentially practical, engineering-linked considerations came much of what is now known as thermodynamics. Tragically, Carnot himself did not live to see any of the far-reaching consequences of his work;- dying from cholera at the early age of 36. However, his work was used and extended by, amongst others, Thomson and

Clausius and, once the problem of reconciling Carnot's work, in which caloric is conserved, with Joule's work demonstrating the interconvertibility of heat and work, had been resolved, modern thermodynamics began to emerge.

The Second Law itself has been stated in various ways but probably the two most common forms are those due to Lord Kelvin (William Thomson) and Rudolf Clausius:

> **Kelvin:**
>
> **It is impossible to transform an amount of heat completely into work in a cyclic process in the absence of other effects.**
>
> **Clausius:**
>
> **It is impossible for heat to be transferred by a cyclic process from a body to one warmer than itself without producing other changes at the same time.**

As may be seen in most thermodynamics' text-books, these two statements of the Second Law were thought to be equivalent. However, it is important to note that they are equivalent for *positive temperatures only*, as will be discussed in Chapter 9. Also, it may be noted that, in chapter 3, the idea of a perpetual motion machine of the first kind was introduced and it was noticed that the existence of such a machine is prohibited by the First Law. It might be noted at this point that the idea of a machine which, in a cyclic process, converts an amount of heat *completely* into work has been suggested also. Such a machine, if permissible, would prove an extremely attractive proposition since it could be used to cool both the deserts and oceans and so provide a huge supply of energy for man's use. However, such a machine is prohibited by the Second Law, as is seen by glancing at the above statement of that law due to Kelvin. These machines, which are still sought by some people, are often called *perpetual motion machines of the second kind*.

These forms of the law are those used at the birth of thermodynamics as a subject in its own right. As mentioned already, the laws were deduced from experiment and observation, and many of the ideas were borrowed from engineering. The notions and experiences of the engineer were used to obtain the laws of heat transformation and it is a tremendous achievement that a theory with many highly abstract concepts should be established by this approach. However, the approach to be adopted here is more mathematical in nature than some earlier arguments and is a modification of the method introduced at the beginning of this century by the mathematician Constantin Carathéodory. Carathéodory became interested in the problem of the formulation of thermodynamics at the instigation of his colleague, the physicist Max Born, and his highly mathematical original paper appeared in 1909. Because of the mathematical complexities of his approach, his work passed largely unnoticed, until the postwar work of such as Buchdahl, Landsberg, Turner and Zemansky made it far more accessible to scientists in general. Here, as a starting point, the Kelvin formulation of the Second Law will be used rather than the more abstract principle of Carathéodory. However, before proceeding with a discussion of the Second Law and its consequences, it is necessary to consider what is meant by a *quasistatic* process but first the meaning of thermodynamic equilibrium must be introduced.

A system in *thermodynamic equilibrium* must satisfy three requirements:

1. *mechanical equilibrium* means there are no unbalanced forces acting on any part of the system or on the system as a whole.

2. *thermal equilibrium* means there are no temperature differences between parts of the system or between the system and its surroundings.

3. *chemical equilibrium* means there are no chemical reactions within the system.

If a system is in thermodynamic equilibrium and the surroundings are kept unchanged, no motion will take place and no work will be done. If the situation is changed so that a finite unbalanced force acts on the system, then the system may proceed from one state to another but, in so doing, *may* pass through non-equilibrium states. A *quasistatic process* is an idealised process during which the system passes only through equilibrium states; that is, it is a process which consists exclusively of a sequence of thermodynamic equilibrium states. Such processes do not occur in nature since a change from one equilibrium state to another is caused usually by outside interference with an existing equilibrium state. However, the interference may be made small and, as a result, the change from one equilibrium state to another may occur very slowly. In this way, approximations to quasistatic processes may be made in actual experiments. Processes which are not quasistatic are, not unnaturally, called **non-static**.

Generalising some earlier results to systems of more than two coordinates, it may be noted that a typical increment of non-thermal energy given in a quasistatic change to its surroundings *by* a system of interest has the form

$$d'W = \sum_i p_i da_i \tag{4.1}$$

where some, or all, of the a's are 'external parameters' or 'generalised deformation coordinates' and the p's are the corresponding generalised forces. (Here, by 'deformation coordinate' is meant a coordinate, such as volume, whose change implies an alteration or deformation in the size and/or shape of the system). Certain 'internal variables' may be included also among the a's but it is found necessary merely that the internal energy, U, and the a's form a complete set of independent variables for the system:

for example, in the case of a gas, a typical increment of quasistatic mechanical work done by the gas is pdV, where dV is the increment of volume swept through by a piston at pressure p.

Now, by (4.1) and (3.3)

$$d'Q = dU + d'W = dU + \sum_i p_i da_i \tag{4.2}$$

is valid for a quasistatic process. This gives an expression for an increment of heat added to a system.

Only infinitesimal changes are required to make a quasistatic process proceed through the same continuum of equilibrium states in reversed time order and so quasistatic processes are *reversible*.

Even the supply of heat to a system may be made quasistatic by using a device known as a **heat reservoir**. This is a very large system at a definite temperature; it is so large, in fact, that it may gain or lose heat without its temperature being sensibly affected. This property may be made exact in the limit of an infinitely large reservoir.

At this point it proves convenient to restrict considerations to a three-coordinate system; that is, a system described by three independent variables. This is done for two reasons:

1.   simple three-dimensional graphs may be used,

2.   all conclusions concerning the mathematical properties of $d'Q$ will hold equally well for systems with more, or less, independent variables.

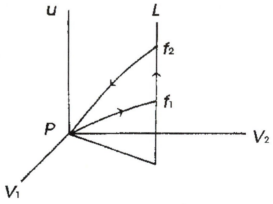

In this figure, the three independent variables are taken to be the internal energy, U, and the two deformation coordinates V_1 and V_2. These are plotted along three rectangular axes and the origin P is an arbitrarily chosen equilibrium state. For a system described by these variables, (4.2) becomes

$$d'Q = dU + p_1 dV_1 + p_2 dV_2 . \qquad (4.3)$$

Suppose that f_1 is an equilibrium state the system may reach from P by a quasistatic adiabatic process. Through f_1 construct a line L, at every point of which the values of V_1 and V_2 are constant; that is, L is a line of constant deformation. Let f_2 be any other equilibrium state on this line. It will be proved now that only one of f_1 and f_2 may be reached from P by a quasistatic adiabatic process - not both!

First it is assumed that both of the paths $P \to f_1$ and $P \to f_2$ are quasistatic adiabatic and it will be shown that this assumption leads to a contradiction of the

Kelvin form of the Second Law. Let the system start at P, proceed to f_1, then to f_2, and finally back to P along the path $f_2 \rightarrow P$. Since the process $P \rightarrow f_2$ is merely a sequence of equilibrium states, the sequence $f_2 \rightarrow P$ may be treated as a quasistatic adiabatic process also. Since f_2 lies above f_1, in passing from f_1 to f_2 the system undergoes an increase in energy at constant V_1 and V_2, during which process no work is done. From the First Law it follows that a quantity of heat Q must be absorbed in the process $f_1 \rightarrow f_2$. However, in the two quasistatic adiabatic processes, no heat is transferred but an amount of work W is done. In the complete cycle $P \rightarrow f_1 \rightarrow f_2 \rightarrow P$ there is no resultant change in energy and so $Q = W$. Hence, the system has performed a cycle in which the sole effect is the absorption of heat and the conversion of this heat completely into work. Since this violates the Kelvin statement of the Second Law, it follows that both f_1 and f_2 may not be reached from P by quasistatic adiabatic processes; only one point, say f_1, on a line of constant V_1 and V_2 (that is, a line of constant deformation) may be reached from P by a quasistatic adiabatic process.

This argument may be repeated for displaced lines L', L'',...... parallel to L and the point f_1 is seen to trace out a surface on which P itself must lie. The complete argument may be repeated for different initial states P and the (U, V_1, V_2) space is found to be decomposed into a family of non-intersecting surfaces. These surfaces may be labelled by a continuously varying parameter ϕ - called the *empirical entropy* - such that distinct values of ϕ refer to distinct surfaces, and conversely. It may be noticed that any two points in a particular surface are connected by quasistatic adiabatic processes but this is *not* so for points in different surfaces.

The method of construction of the so-called level surfaces implies that each quasistatic adiabatic process lies in one such surface. Now use is made of the fact that the decomposition of the space into a family of level surfaces is equivalent mathematically to saying that there exists a function λ (U, V_1, V_2) of the thermodynamic variables such that, for quasistatic processes, $d'Q = \lambda d\phi$, where $d\phi$ is the exact differential of the empirical entropy. This may be seen as follows:

Consider a family of surfaces

$$\phi\ (y_1, y_2, y_3) = c.$$

Such a family has been obtained for the case in question and for each value of c, within some continuous range, one surface is selected. For each increment of a line in such a surface

$$d'Q = \sum_{j=1}^{3} X_j\ (y_1, y_2, y_3)dy_j = 0$$

where the expression for $d'Q$ is just an alternative way of writing (4.3); in other words, the notation has been changed for convenience because of that which follows but the two are linked as indicated below:

$$dU \rightarrow dy_1, 1 \rightarrow X_1; dV_1 \rightarrow dy_2, p_1 \rightarrow X_2; dV_2 \rightarrow dy_3, p_2 \rightarrow X_3.$$

However, for distinct surfaces, $d'Q$ does not equal zero; that is , for a line element connecting two infinitesimally close surfaces $d'Q$ does not equal zero.

Any element *ds* of a line describing a quasistatic change satisfies two equations. Write

$$ds = (dy_1, dy_2, dy_3), \quad R = (X_1, X_2, X_3).$$

Then, from the condition for a quasistatic adiabatic change

$$d'Q = R.ds = 0.$$

Also, due to the existence of a family of level surfaces

$$R'.ds = 0$$

where $R' = \left(\partial\phi / \partial y_1, \partial\phi / \partial y_2, \partial\phi / \partial y_3 \right)$.

This result follows easily on differentiating $\phi (y_1, y_2, y_3) = c.$

Hence, at *each point P* in the space, any element *ds* of a line in a surface $\phi (y_1, y_2, y_3) = c$ is orthogonal to two vectors, R and R' , whose directions are functions of the coordinates of P. It follows that, throughout the simultaneous range of definition of the vectors R and R' , these are parallel vectors and they are perpendicular to the element of surface at P. Therefore, there exists a scalar function of position , λ , such that

$$R = \lambda R'.$$

Then

$$d'Q = R.ds = \lambda R'.ds = \lambda d\phi.$$

It should be noted that, for quasistatic adiabatic processes, $d\phi = 0$ since such processes are confined to surfaces of constant ϕ but, of course, this is not true of processes in general.

According to equation (4.3), the three independent thermodynamic variables are the internal energy and two deformation coordinates V_1 and V_2. It must be shown next that the empirical temperature, t, depends on at least one of V_1 and V_2. Certainly t must depend on at least one of the three variables mentioned; otherwise there would be *four* independent variables for the three dimensional space described. Hence, suppose t to be a function of the internal energy, U, alone. If this is the case, arbitrary variations in the remaining variables will leave t and, therefore, the internal energy unaltered. Such arbitrary variations are not restricted if the system is enclosed adiabatically. However, if this is the case, it implies that both $d'Q$ and dU equal zero in the equation

$$d'Q = dU - d'W$$

for all changes of V_1 and V_2. Hence, by the First Law, $d'W$ must be zero also for all such changes. This implies the existence of a relation between the supposedly independent V_1 and V_2, which is contrary to hypothesis. Therefore, any empirical temperature must depend on at least one of the variables V_1 and V_2. It will be assumed that the empirical temperature does, in fact, depend on V_1 and, in expressions involving V_1, the empirical temperature may be introduced by virtue of this relation.

Now write $d'Q = \lambda d\phi$, where λ is an integrating factor and $d\phi$ an exact differential. Then

$$d'Q = dU + p_1 dV_1 + p_2 dV_2$$

$$= \lambda \frac{\partial \phi}{\partial U} dU + \lambda \frac{\partial \phi}{\partial V_1} dV_1 + \lambda \frac{\partial \phi}{\partial V_2} dV_2 .$$

Hence,

$$\lambda \frac{\partial \phi}{\partial U} = 1; \lambda \frac{\partial \phi}{\partial V_1} = p_1; \lambda \frac{\partial \phi}{\partial V_2} = p_2 .$$

This equation shows that $\partial \phi / \partial U$ is non-zero; and so ϕ may be used instead of U as one of the independent thermodynamic variables.

Now consider the thermal equilibrium of two systems, each described by three independent variables. It might be noted at this point that attention has been confined to systems described by just three independent variables since, as mentioned earlier, any results obtained for such systems will apply also to systems described by more than three independent variables. Again, in what follows, the two systems could be described by different numbers of independent variables but, here, attention will be restricted to the simplest case; - although the extension to higher numbers of independent variables is quite easy and straightforward once the technique involved is grasped. Hence, suppose the variables corresponding to the two systems are

$$t', \phi', V_2' \text{ and } t'', \phi'', V_2''.$$

Thermal equilibrium means that $t' = t'' = t$, say. As has been shown already, an integrating factor exists for each system. This result may be applied to each system separately and also to the joint system. Using

$$d'Q = d'Q' + d'Q''$$

for the increment of thermal energy supplied to the combined system by quasistatic processes, it follows that

$$\lambda d\phi = \lambda' d\phi' + \lambda'' d\phi''.$$

Also, $d\phi$ may be expressed in terms of all the independent variables used to describe the composite system:

$$d\phi = (\partial\phi/\partial t)dt + (\partial\phi/\partial\phi')d\phi' + (\partial\phi/\partial\phi'')d\phi''$$
$$+ (\partial\phi/\partial V_2')dV_2' + (\partial\phi/\partial V'')dV''_2.$$

From these latter two equations, it follows that

$$\partial\phi/\partial\phi' = \lambda'/\lambda \; ; \quad \partial\phi/\partial\phi'' = \lambda''/\lambda \; ;$$

and

$$\partial\phi/\partial t = \partial\phi/\partial V_2' = \partial\phi/\partial V_2'' = 0.$$

Then,

$$\partial^2\phi/\partial\phi'\partial t = \partial^2\phi/\partial\phi''\partial t = 0$$

that is

$$\frac{\partial}{\partial t}\left(\frac{\lambda'}{\lambda}\right) = \frac{\partial}{\partial t}\left(\frac{\lambda''}{\lambda}\right) = 0.$$

As was first noted by Born, this implies

$$\frac{1}{\lambda'}\frac{\partial\lambda'}{\partial t} = \frac{1}{\lambda''}\frac{\partial\lambda''}{\partial t} = \frac{1}{\lambda}\frac{\partial\lambda}{\partial t}.$$

Here λ' depends on t and on variables associated with the first system; whereas λ'' depends on t and variables associated with the second system. Hence, each expression in this last equation must be a function of t only, $g(t)$ say. $g(t)$ is a universal function of the chosen empirical temperature scale. If a positive function λ is an integrating factor, then a corresponding negative function $-\lambda$ is an integrating factor also and both choices lead to the same function $g(t)$. No special assumptions will be made concerning the sign of λ.

Integration shows that the integrating factor has the form

$$\lambda'\left(t,\phi',V_2'\right) = \Phi'\left(\phi',V_2'\right)\exp\int_{t_0}^{t} g(x)dx$$

or

$$\lambda''\left(t,\phi'',V_2''\right) = \Phi''\left(\phi'',V_2''\right)\exp\int_{t_0}^{t} g(x)dx.$$

Here t_0 is a standard empirical temperature which is assumed to be the same for all physical systems. At $t = t_0$,

$$\lambda'\left(t_0,\phi',V_2'\right) = \Phi'\left(\phi',V_2'\right) \quad \text{and} \quad \lambda''\left(t_0,\phi'',V_2''\right) = \Phi''\left(\phi'',V_2''\right).$$

It is seen that the integrating factor for the combined system may be written in this form also.

If now, not only t_0 , but also a constant C are given for a whole class of physical systems, then

$$d'Q \;=\; \lambda d\phi \;=\; TdS,$$

where

$$T(t) \;=\; C \exp \int_{t_0}^{t} g(x)dx \tag{4.4}$$

and

$$S(\phi) \;=\; C^{-1} \int_{\alpha}^{\phi} \Phi(\phi, V_2', V_2'')dy. \tag{4.5}$$

Here α refers to a standard value of ϕ for the particular system under discussion, and the value $\phi = \alpha$ has the property that $S(\alpha) = 0$. T is called the *absolute temperature* and depends only on t_0 and the empirical temperature t, while S is called the *entropy* and depends on the variables V_2' , V_2'', the function ϕ and the value $\phi = \alpha$.

Therefore, if the process under consideration is *quasistatic,* it is seen that the equation

$$d'Q \;=\; TdS$$

holds. This is one of the most important equations occurring in thermodynamics; indeed, it is one of the most important equations in physics!

At this point, it does not seem unreasonable to ask "What is entropy?". Most people feel they have some intuitive idea concerning what is meant by temperature, heat, flow of heat, and many other concepts which, although met in everyday life, are seen to be parts of thermodynamics. Entropy seems different. Whichever approach is adopted, the introduction of entropy appears to be due to a purely mathematical manipulation. However, entropy is not simply a hazy, mathematical concept; it is, rather, a measurable, physical quantity - for example, when a substance is taken from one state to another by a series of small, quasistatic steps, the increase of entropy is found by dividing the quantity of heat supplied in each of these steps by the temperature at which it was supplied and then summing all these small contributions.

When the absolute temperature has been defined by (4.4), one question remains; - what form does the equation of an ideal gas take in terms of T?

In terms of t, the equation of an ideal gas is

$$pV \;=\; Rt$$

as was mentioned in Chapter 3. Also, from the Joule - Kelvin experiment, it is known that the internal energy is independent of volume; that is, $U = U(t)$.

Therefore,

$$d'Q = dU + pdV$$

$$= \frac{dU}{dt}dt + \frac{t}{V}dV$$

$$= t\left\{\frac{1}{t}\frac{dU}{dt}dt + d(\log V)\right\}$$

or, if χ is defined by

$$\log \chi = \int \frac{1}{t}\frac{dU}{dt}dt$$

then

$$d'Q = td(\log \chi\, V).$$

Therefore, $1/t$ may be chosen as integrating factor: that is, if we choose $t = \lambda$ and $\log \chi\, V = \phi$, the latter equation becomes

$$d'Q = \lambda d\phi\,.$$

Also, it has been shown that

$$g(t) = \frac{\partial \log \lambda}{\partial t}$$

is a universal function; the same however the integrating factor is chosen. From this, it follows that

$$\lambda = \exp \int_{t_0}^{t} g(x)dx\,.$$

Hence,

$$T = C\exp \int_{t_0}^{t} g(x)dx = C\lambda = Ct.$$

Therefore,

$$pV = RT/C.$$

Hence, absolute temperature is proportional to empirical temperature.

Using the Kelvin form of the Second Law, the equation $d'Q = TdS$ has been derived. The actual derivation is rather long and involved. However, it might be noted that the derivation may be split into three distinct parts:

1.   showing that the thermodynamic phase space may be decomposed into a family of level surfaces which may be labelled by a continuously varying parameter ϕ,

2.   showing that a consequence of (1) is the existence of an integrating factor for the inexact differential $d'Q$,

3.   proceeding to introduce the absolute temperature, T, and the entropy, S.

Exercises C

In the following examples, the Second Law will be required. The changes may be assumed quasistatic and, for such changes,

$$d'Q = TdS.$$

(1)   The following thermodynamic functions may be defined:

Helmholtz free energy, $F = U - TS$

Enthalpy, $H = U + pV$

Gibbs free energy, $G = U + pV - TS.$

Show that

$$dH = TdS + Vdp$$

and obtain analogous results for dF and dG.

(2)   By considering the exact differentials dF, dH, dG and dU, derive the so-called Maxwell relations:

$$\left(\frac{\partial T}{\partial V}\right)_S = -\left(\frac{\partial p}{\partial S}\right)_V ; \left(\frac{\partial T}{\partial p}\right)_S = \left(\frac{\partial V}{\partial S}\right)_p$$

$$\left(\frac{\partial T}{\partial V}\right)_p = -\left(\frac{\partial p}{\partial S}\right)_T ; \left(\frac{\partial T}{\partial p}\right)_V = \left(\frac{\partial V}{\partial S}\right)_T .$$

(3)   For a paramagnetic system, which is kept at constant volume and pressure, the term $(-pdV)$ applicable to a fluid is replaced by the work term BdM, where B is the magnetic field and M the magnetisation which is assumed parallel to H. Obtain the analogues of the Maxwell relations and show that the heat capacities satisfy

$$C_B = T\left(\frac{\partial S}{\partial T}\right)_B ; C_M = T\left(\frac{\partial S}{\partial T}\right)_M$$

$$C_B - C_M = -T\left(\frac{\partial M}{\partial T}\right)_B \left(\frac{\partial B}{\partial T}\right)_M = T\left(\frac{\partial M}{\partial T}\right)_B^2 \left(\frac{\partial B}{\partial M}\right)_T .$$

(4)   For a fluid, prove that

$$\left(\frac{\partial U}{\partial V}\right)_T = T\left(\frac{\partial S}{\partial V}\right)_T - p = T\left(\frac{\partial p}{\partial T}\right)_V - p$$

Hence show that Joule's Law implies the existence of a function of volume only such that

$$pg(V) = T.$$

If Joule's Law holds as well as the equation of state

$$pV = f(T),$$

find the form of the functions f and g.

(5)   With the usual meaning for the various symbols, show that

$$C_p - C_V = T\left(\frac{\partial V}{\partial T}\right)_p \left(\frac{\partial p}{\partial T}\right)_V = \frac{TV\alpha_p^2}{\kappa_T}.$$

(6)   By considering

$$TdS = dU + pdV,$$

show that

$$\left(\frac{\partial U}{\partial V}\right)_T = T\left(\frac{\partial S}{\partial V}\right)_T - p.$$

Hence show that the internal energy ,U, of a classical ideal gas, for which $pV = AT$, is independent of volume. If, in addition, the heat capacity C_V is a constant, show that

$$U(T) - U(0) = C_V T.$$

A system to which the above assumptions apply is enclosed in an upright, insulated cylinder by the weight of a frictionless, insulated piston. It is in equilibrium at pressure p_1, volume V_1 and absolute temperature T_1. Upon gently placing a weight on the piston, a new equilibrium state, with pressure p_2, volume V_2 and temperature T_2, is attained. Show that the increase in the the internal energy of the gas is $p_2(V_1 - V_2)$ and deduce that

$$\frac{T_2}{T_1} = \frac{C_V + \lambda A}{C_V + A},$$

where $\lambda = p_2/p_1$.

5

The Second Law and Non-static Processes.

In the previous chapter, it was seen that, provided the process under consideration is quasistatic, the equation

$$d'Q = TdS$$

holds. It is necessary now to discuss briefly the status of non-static processes.

Referring back to the decomposition of what may be termed thermodynamic phase space into a family of level surfaces, it may be remembered that any two points in such a surface may be linked by a quasistatic adiabatic process. Also, any two points (or states) may be linked theoretically by an adiabatic process since all this means is that a system may proceed from one state to another without exchange of heat. However, if the two points (or states) lie in different level surfaces, the adiabatic process linking them cannot be quasistatic since this would imply $dS = 0$ for every increment of such a process. Hence, if the process is adiabatic, it must be non-static. To sum up, points (or states) with different entropy values cannot be linked by any quasistatic adiabatic process - only by a non-static adiabatic process. Of course, such states may be linked by a quasistatic process but, if it is quasistatic, it cannot be adiabatic - some exchange of heat with the surroundings must be involved.

Consider a system possessing three independent variables T, V_1 and V_2 and let this system be taken around the cycle illustrated in the figure.

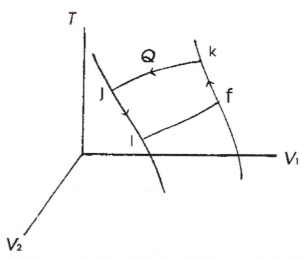

Suppose the initial state of this system is i and suppose it undergoes a non-static adiabatic process to a state f, where i and f are both assumed to be equilibrium states of the system. Then, the entropy change is

$$\Delta S = S_f - S_i.$$

During this process, a temperature change may, or may not, have occurred. Whether it has or not, now suppose the system undergoes a quasistatic adiabatic process $f \rightarrow k$ to bring its temperature to that of some arbitrary heat reservoir at temperature T'. Since S_f and S_k are equal,

$$\Delta S = S_k - S_i.$$

The system may be brought into contact with the reservoir and caused to undergo an isothermal process $k \rightarrow j$ until its entropy is the same as it was initially. A quasistatic adiabatic process $j \rightarrow i$ returns the system to its initial state and, since S_j and S_i are equal.

$$\Delta S = S_k - S_j.$$

The only heat transfer, Q, that has taken place in the cycle is during the isothermal process where

$$Q = T'(S_j - S_k).$$

Also, a net amount of work, W, has been done in the cycle where

$$W = Q.$$

From the Second Law, it is clear that the heat Q cannot have entered the system - that is, Q cannot be positive - for then, the cyclic process would have been such as to produce no effect other than the extraction of heat from a reservoir and the performance of an equivalent amount of work.

Hence,

$$Q \leq 0,$$

from which it follows that

$$T'(S_j - S_k) \leq 0$$

or

$$\Delta S = S_k - S_j \geq 0.$$

Here it has been assumed that an entropy change is associated with the original non-static adiabatic process. If this were not so, it would be possible to return the system to state i by one quasistatic adiabatic process. Since the net heat transferred in this cycle is zero, the net work would be zero also. Under these circumstances, the system and its surroundings would have been restored to their initial states without producing changes elsewhere - implying that the original process was quasistatic. This is contrary to the original assertion, and so the entropy of the system cannot remain unchanged.

Again, the system considered was assumed homogeneous and of uniform temperature and pressure. If this were not so, it would be necessary to subdivide the system into parts - each one infinitesimal in an extreme case - and to ascribe a definite temperature and pressure to each part, so that each part would have a definite entropy depending on its coordinates. The entropy of the system as a whole would be defined to be the sum of the entropies of the various parts. If it is possible to return each part to its initial state in the manner described earlier, using the same reservoir for each part, it follows that ΔS is positive for the whole system.

The final result is that the entropy of a system in a given state cannot be decreased adiabatically for a thermodynamics in which the absolute temperature is positive and heat tends to flow from high to low absolute temperatures. This is a statement of the **principle of the increase of entropy** of systems in adiabatic enclosures.

Suppose the transfer of an increment of heat $d'Q$ to a system, whose absolute temperature is T, is accompanied by quasistatic or non-static adiabatic disturbances of the system. The resultant process is more general than those considered so far, since it is neither necessarily quasistatic nor adiabatic.

The relation

$$d'Q = TdS$$

is replaced by an inequality, which may be obtained from the above principle

$$d'Q \leq TdS \tag{5.1}$$

where the equality sign holds if the process is quasistatic.

Relation (5.1) is the basic inequality of thermodynamics. It remains valid for negative $d'Q$ when heat is given up by the system. It holds for any system undergoing any infinitesimal process, *provided* the system does not exchange mass with its surroundings; - that is, it is a **closed** system. Although the process considered has been subdivided into an exchange of heat with the surroundings and an adiabatic process, clearly any process may be supposed divided in this way and so no loss of generality has been incurred.

6

The Third Law.

The Third Law states that

The contribution to the entropy of a system by each aspect which is in internal thermodynamic equilibrium tends to zero at the absolute zero of temperature.

One of the most important generalisations which may be made from this law is that *the specific heats of all substances tend to zero at absolute zero*. This statement will not be discussed further here but it is known experimentally that all specific heats do tend to zero at the absolute zero of temperature. The vanishing of the specific heats is of great importance since it allows the use of absolute zero as a reference level for all thermodynamic calculations. As has been seen already, the definition of entropy is based on the Second Law but is sufficient to define differences of entropy only. Thus, if S_T and S_0 are the entropies of a substance in thermodynamic equilibrium at temperatures T and zero respectively,

$$S_T - S_0 = \int_0^T \frac{CdT}{T}.$$

Since the specific heat tends to zero at absolute zero, this integral is finite. Hence, by stating that S_0 is zero for all simple substances, the Third Law allows a unique value to be allotted to the entropy at any temperature:

$$S_T = \int_0^T \frac{CdT}{T}.$$

Strictly speaking, the accurate statement of the Third Law is that given above which, stated more simply, is that the entropy of every system at absolute zero may be taken equal to zero. However, another commonly used form is the so-called **unattainability principle**, which states that **it is impossible to cool any substance to the absolute zero of temperature.** It is interesting to note that this statement has never been violated experimentally, although, as long ago as 1960, by using nuclear cooling techniques, Kurti did succeed in reaching a temperature of 10^{-6} degrees, and even lower temperatures have been recorded more recently.

The question of whether the principle of unattainability implies the vanishing of the entropy at absolute zero must be examined now. Consider a quasistatic adiabatic change between two states of a system brought about by varying some external parameter from a value α to a value β. The system passes from a state with temperature T_1 and entropy S_1^α to a state with temperature T_2 and entropy S_2^β.

Using the Second Law,

$$S_1^\alpha = S^\alpha(0) + \int_0^{T_1} \frac{C_\alpha}{T} dT,$$

$$S_2^\beta = S^\beta(0) + \int_0^{T_2} \frac{C_\beta}{T} dT,$$

where $S^\alpha(0)$ and $S^\beta(0)$ are the entropies at absolute zero. For a quasistatic adiabatic change, $S_2^\beta = S_1^\alpha$, so that

$$S^\beta(0) + \int_0^{T_2} \frac{C_\beta}{T} dT = S^\alpha(0) + \int_0^{T_1} \frac{C_\alpha}{T} dT.$$

Thus, if T_2 is to be zero,

$$\int_0^{T_1} \frac{C_\alpha}{T} dT = S^\beta(0) - S^\alpha(0).$$

This equation gives the value of T_1 which will lead to an end temperature of absolute zero. It has a real solution provided the right-hand side is positive. However, if absolute zero is unattainable from any temperature T_1

$$S^\alpha(0) \geq S^\beta(0).$$

If the same transition is used now in the reverse direction to reach a temperature $T_1 = 0$ from an initial temperature T_2, the argument would lead to the inequality

$$S^\alpha(0) \leq S^\beta(0).$$

Hence, it follows that

$$S^\alpha(0) = S^\beta(0).$$

The restriction on a system implied by the unattainability principle is that entropy *differences* between different states of the system disappear at the absolute zero of temperature. It is *not* necessary for the entropies themselves to vanish.

7

Extension to open and non-equilibrium systems.

As was mentioned earlier, the simplest thermodynamic systems are two-coordinate systems; that is, they have two independent variables. For example, a gas may be considered and the internal energy, U, and volume, V, taken as independent variables, then, for any increment of a quasistatic process,

$$TdS \;=\; dU \;+\; pdV,$$

where p is the pressure, holds. If the system is split up into various parts, this equation may be generalised to allow for several pressures p_i and external parameters V_i , since the pressure need not be uniform throughout the volume of the gas. If there are r external parameters, let $\boldsymbol{p}$ and $\boldsymbol{dV}$ represent the appropriate r-dimensional vectors so that

$$TdS \;=\; dU \;+\; \boldsymbol{p.dV}.$$

However, as follows immediately from Chapter 5 and using the First Law also, for an increment of some general process

$$TdS \;\geq\; dU \;+\; pdV \tag{7.1}$$

where equation (5.1) has been used and where the equality sign holds if the process involved is quasistatic.

The increments of the processes considered in equation (7.1) have equilibrium states as end - points and, so far, systems under consideration have been supposed *closed* - that is, unable to exchange matter with their surroundings. However, at times it becomes necessary to consider *open* systems, which may exchange matter with their surroundings, and increments of processes whose end -points are non-equilibrium states. One way of generalising previous results to cover this new situation is to consider an r - coordinate, χ - component simple thermodynamic system. A state of such a system may be specified uniquely in terms of an internal energy function, U, r external parameters $V_1 ,V_2 ,.......,V_r$, and χ additional independent variables $x_1, x_2,......,x_\chi$ such that the entropy of the system is defined to satisfy

$$S\big(aU, aV_1,....., aV_r, ax_1,....., ax_\chi \big) = aS\big(U, V_1,....., V_r, x_1,....., x_\chi \big) \tag{7.2}$$

for all positive a.

At this stage, the new variables x_i may be thought of as supplementing the information provided by knowledge of the internal energy and of the external parameters (or deformation coordinates) V_i ; that is, they may be regarded as providing additional information about the internal state of the system.

Any thermodynamic function f which may be expressed in terms of a complete set of independent thermodynamic variables X_1, X_2,.... such that

$$f(aX_1, aX_2,......) = af(X_1, X_2,......)$$

is called an **extensive** variable in thermodynamics; mathematically, it is a homogeneous function of degree one.

If the relation

$$f_0 = f(X_1, X_2,....)$$

may be solved for X_i, suppose the solution is

$$X_i = g_i (X_1,......,X_{i-1}, f_0, X_{i+1},....).$$

Then the solution of

$$af_0 = f(aX_1, aX_2,......)$$

will be

$$aX_i = g_i (aX_1,...., aX_{i-1}, af_0, aX_{i+1},....)$$

and, from these latter two equations, it follows that X_i is also an extensive variable. Hence, it follows from (7.2) that $S, U, V_1,....,V_r, x_1,....,x_\chi$ are **all** extensive variables.

This introduction to the idea of extensivity is somewhat abstract and it is not clear what it means in simple physical terms. Hence, consider two identical systems and suppose they are taken together as a single system. The value of the volume for the composite system is twice that for a single subsystem; and similarly for the value of the total number of particles - in the composite system, its value is twice that for a subsystem. Parameters which have values in a composite system equal to the sum of the values in each of the subsystems - that is, parameters which depend on the *extent* of the system - are called *extensive parameters*. Parameters which are not extensive are said to be *intensive*. There can be little doubt that knowledge of these two terms is not vital to an understanding of thermodynamics - both are merely pieces of jargon - but the words do occur fairly frequently in the literature surrounding the subject, so having met them can prove useful.

For a simple, one-coordinate, one component system, equation (7.2) becomes

$$S(aU, aV, aN) = aS(U, V, N). \tag{7.3}$$

This equation refers to a system consisting of one type of particle only; V represents volume and N number of particles. Also, as usual, U represents the internal energy.

Now, for such a system,

$$TdS = T\left(\frac{\partial S}{\partial U}\right)_{V,N} dU + T\left(\frac{\partial S}{\partial V}\right)_{N,U} dV + T\left(\frac{\partial S}{\partial N}\right)_{U,V} dN. \tag{7.4}$$

It is **convention** to use the following definitions:

$$T\left(\frac{\partial S}{\partial U}\right)_{V,N} = 1, \, T\left(\frac{\partial S}{\partial V}\right)_{N,U} = p, \, T\left(\frac{\partial S}{\partial N}\right)_{U,V} = -\mu \, , \tag{7.5}$$

where μ is called the *chemical potential* and, as usual p is the pressure.

These definitions may be seen to be correct dimensionally and, using them, equation (7.4) becomes

$$TdS = dU + pdV - \mu dN \, . \tag{7.6}$$

This equation applies to the general case of varying particle number. In the special case when the number of particles is constant ($dN = 0$), equation (7.5) takes the form

$$TdS = dU + pdV.$$

However, it is important to realise that the derivation of equation (7.6) given here is a purely mathematical one based on the definition of the entropy introduced in equation (7.2). It is found in practice that equations such as (7.6) do, in fact, apply in genuine physical situations and it is that, and to a certain extent, that alone, which justifies equation (7.6) as derived in the approach adopted here. In general, it is important to realise that, beautiful as some mathematical deductions may be, their relevance in physics is determined solely by their success in describing the physical events, they purport to describe, accurately

If equation (7.6) is written in the form

$$dU = TdS - pdV + \mu dN \, ,$$

the pairs of variables $(S,T), (V,p), (N,\mu)$ are seen to play mathematically analogous roles apart from sign. These pairs - each consisting of one extensive and one intensive variable - are referred to as *canonically conjugate thermodynamic variables*. On occasions, however, it proves convenient to consider U and T, rather than S and T, as conjugate variables. Again, this terminology is really only jargon, and noting which pairs of variables are said to be canonically conjugate does not assist dramatically towards a deeper understanding of thermodynamics. Nevertheless, this terminology does occur in the literature and so, an acquaintance with it is worthwhile.

The increment

$$d'W = pdV - \mu dN$$

is referred to as the generalised increment of work done by the system. This concept of generalised work done includes the work, μdN , which is, in a sense, internal to the system.

The most important properties of simple thermodynamic systems may be established now. However, for simplicity, attention will be confined to a simple, one-coordinate, one-component system and so, equation (7.3) will be the starting point for the manipulations. Hence, first differentiate (7.3) with respect to a to give

$$S = \left(\frac{\partial S}{\partial(aU)}\right)_{V,N} U + \left(\frac{\partial S}{\partial(aV)}\right)_{N,U} V + \left(\frac{\partial S}{\partial(aN)}\right)_{U,V} N .$$

Since (7.3) holds for all a, it must hold for $a = 1$ and so, the above equation becomes

$$S = \frac{1}{T}(U + pV - \mu N)$$

or

$$TS = U + pV - \mu N , \tag{7.7}$$

where the definitions (7.5) have been used.

Equation (7.7) is the **Euler relation** which links the extensive and intensive variables of a system.

Again, from equation (7.7), it is seen that

$$TdS + SdT = dU + pdV + Vdp - \mu dN - Nd\mu .$$

Subtracting equation (7.6) from this result gives

$$SdT = Vdp - Nd\mu$$

or, more usually,

$$SdT - Vdp + Nd\mu = 0 . \tag{7.8}$$

This is the well-known **Gibbs-Duhem relation** which can prove extremely useful in a number of thermodynamic manipulations. It is important to note that this relation shows that only two of the three intensive variables are independent.

From equations (7.1) and (7.6), it is seen that, for a closed simple system

$$dU + pdV - TdS = \mu dN \leq 0. \tag{7.9}$$

Some special cases of this equation are listed :

Condition of system or type of process.	Equation valid for the process.	Inequality (7.9) valid for the process.
(a) adiabatic	$d'Q = dU + pdV$ $=0$	$dS \geq 0$
(b) adiabatically and mechanically isolated	$dU = pdV = 0$	$dS \geq 0$
(c) isothermal	$dT = 0$	$d(U\text{-}TS) + pdV \leq 0$
(d) isothermal, mechanically isolated	$dT = 0,$ $pdV = 0$	$d(U - TS) \leq 0$
(e) isentropic	$dS = 0$	$dU + pdV \leq 0$
(f) isentropic, mechanically isolated	$dS = 0,$ $pdV = 0$	$dU \leq 0$

Now apply the transformation $X_i \rightarrow X_i' = aX_i$ to the extensive variables U, V, N in equation (7.3) so that the entropy undergoes the same transformation. Then

$$\frac{1}{T'} = \left(\frac{\partial S'}{\partial U'}\right)_{V',N'} = \left(\frac{\partial S}{\partial U}\right)_{V,N} = \frac{1}{T}.$$

Hence, T is seen to be an intensive variable and, by analogous arguments, p and μ are seen to be intensive variables also. Therefore, as mentioned previously, **all** the variables appearing in equations such as (7.6), (7.7), (7.8) and (7.9) are either extensive or intensive.

Finally, consider generalised forms of the Euler relation (7.7) and the Gibbs - Duhem relation (7.8). In this instance, consider a simple thermodynamic system described by n extensive variables, denoted by E_j , and $(n\text{-}1)$ intensive variables, denoted by I_j , so that the equations become

$$\sum_{j=1}^{n} E_j I_j = 0 \tag{7.10}$$

where $I_1 = 1$ and

$$\sum_{j=1}^{n-1} E_j dI_j = 0 \tag{7.11}$$

respectively.

In the Euler equation (7.7) and the Gibbs-Duhem equation (7.8), the only non-zero E_j's and I_j's were , apart from sign, $E_1 = U, E_2 = S, E_3 = V, E_4 = N$ and $I_1 = 1$, $I_2 = T, I_3 = p, I_4 = \mu$.

Now differentiate (7.10) with respect to E_r keeping I_i , for $i \neq k$, constant to give

$$\sum_{j=1}^{n} I_j \left(\frac{\partial E_j}{\partial E_r} \right)_X + E_k \left(\frac{\partial I_k}{\partial E_r} \right)_X = 0 \qquad (7.12)$$

where $X = \{I_j \; ; \; i \neq k\}$

However, from (7.11),

$$\left(\frac{\partial I_k}{\partial E_r} \right)_X = 0$$

and so, (7.12) becomes

$$\sum_{j \neq r} I_j \left(\frac{\partial E_j}{\partial E_r} \right)_X + I_r = 0 \, .$$

Using (7.10), this equation may be written

$$\sum_{j \neq r} \left\{ \left(\frac{\partial E_j}{\partial E_r} \right)_X - \frac{E_j}{E_r} \right\} I_j = 0 \, .$$

In this equation, which holds for all values of r, the I_j are independent, and so

$$\left(\frac{\partial E_j}{\partial E_r} \right)_X = \frac{E_j}{E_r} \, .$$

This relation, which is seen to be a direct consequence of the Gibbs-Duhem relation, is useful for simplifying thermodynamic expressions and in a variety of thermodynamic manipulations. Also, it might be noted that , because of the Gibbs-Duhem relation, keeping all but one of the intensive variables constant, means **all** the intensive variables are kept constant. For a simple, one-coordinate, one-component system, such as considered earlier, the extensive variables are U, S, V, N and the corresponding intensive variables are $1, T, p, \mu$. The above result leads, in this case, to such specific conclusions as

$$\left(\frac{\partial N}{\partial V} \right)_{T,p} = \left(\frac{\partial N}{\partial V} \right)_{p,\mu} = \left(\frac{\partial N}{\partial V} \right)_{\mu,T} = \frac{N}{V} \, .$$

8

Thermodynamic cycles.

In the conventional expositions of thermodynamics, the Carnot cycle plays an important role in the arguments which lead to the deduction of the existence of an entropy function and an absolute temperature. Here the existence of these functions has been deduced already and so it suffices to define the Carnot cycle and to investigate its various properties.

Before proceeding further with this discussion, it is worth examining very briefly the role played by the Carnot cycle in the engineering method for deducing the consequences of the Second Law. Again, this method takes as its starting point the Kelvin statement of the Second Law. A particularly simple reversible cycle, called the Carnot cycle, is defined and it is proved that an engine operating in this cycle between reservoirs at two different temperatures is more efficient than any other engine operating between the same two reservoirs. After proving that all Carnot engines operating between the same two reservoirs have the same efficiency, regardless of the substance undergoing the cycle, the absolute temperature is defined. The so-called **Clausius theorem** is derived and, from it, the existence of the entropy function. This method is both rigorous and general. If interest is focussed on the design and manufacture of heat engines and refrigerators , it is essential to employ principles that hold regardless of the nature of the materials involved. However, if interest is focussed on the behaviour of systems, their coordinates, their equations of state, their properties, their processes, etc., apart from their use in the cylinders of engines and refrigerators, then it is desirable to adopt the method used here which is more closely associated with the coordinates and equations of actual systems.

A Carnot engine is an engine whose working fluid has the following properties:

(a)   in each cycle, it passes through the same physical (equilibrium and non-equilibrium) states;

(b)   it consists of one and the same phase and it has a fixed mass throughout a cycle;

(c)   it withdraws an amount of heat Q_1 from a large heat reservoir at temperature T_1 and delivers an amount of heat Q_2 to a similar reservoir at temperature T_2 , with $T_2 < T_1$, in every cycle;

(d)   it exchanges no other heat with its surroundings and there are no other entropy changes in the engine.

If the reservoirs are included in the system, then the engine thus specified may be regarded as an isolated system and so its entropy may not decrease in a cycle. By (a), the working fluid returns to its initial state after each cycle and so does not contribute an entropy change. If the reservoirs are large enough for their temperatures to remain constant, they will contribute an entropy change

$$-\frac{Q_1}{T_1} + \frac{Q_2}{T_2} \geq 0 \qquad (8.1)$$

by virtue of (c) and (d). Using (b) and the First Law, it is seen that the working fluid performs an amount of work W **on** its surroundings which is given by

$$Q_1 \;=\; W + Q_2.$$

The efficiency, η , of such an engine may be defined as the ratio of work done to heat added per cycle. Hence,

$$\eta = \frac{W}{Q_1} = \frac{Q_1 - Q_2}{Q_1}\;. \qquad (8.2)$$

It may be noted from (8.1) that

$$Q_1 / Q_2 \leq T_1 / T_2$$

and so

$$\eta \leq \frac{T_1 - T_2}{T_1}\;.$$

Hence, the upper limit to the efficiency is independent of the amounts of heat Q_1 and Q_2 and, therefore, of the nature of the working fluid; and the efficiency itself is independent of Q_1 and Q_2 if the Carnot cycle is performed quasistatically. Also, a non-static cycle, performed between temperatures T_1 and T_2 , is less efficient than a quasistatic cycle performed between the same temperatures. By appealing to the unattainability of the absolute zero of temperature, the stipulated inequality $T_1 > T_2$ may be extended to

$$T_1 > T_2 > 0$$

and so

$$\eta \leq \frac{T_1 - T_2}{T_1} < 1.$$

In the original discussion of the Carnot engine given here, the word *reversible* was introduced. In the presentation of the general topic *thermodynamics* as given here: however, this term does not appear. However, the question of what happens if a Carnot engine is run in the reverse direction may be answered easily without recourse to the definition of reversibility.

Consider a Carnot engine, for which equation (8.1) holds, run in reverse so that the heat Q_2 is absorbed at temperature T_2 and the heat Q_1 is rejected at temperature T_1. In this case, equation (8.1) becomes

$$\frac{Q_1}{T_1} - \frac{Q_2}{T_2} \geq 0$$

that is,

$$\frac{Q_1}{Q_2} \geq \frac{T_1}{T_2}.$$

The quantity W introduced above may be interpreted, in this case, as the work done **on** the working fluid for the reversed cycle. This work has to be done to enable the reversed engine to absorb heat **from** the colder reservoir. The reversed engine effectively acts like a refrigerator, while, in its forward cycle, the engine may be thought of as designed primarily to produce mechanical work. For its forward cycle, it is desirable for the efficiency, as defined here, to be as large as possible and an upper bound is found for it. However, for the reversed cycle, it is desirable for the efficiency to be as small as possible and a lower bound is found in this case. For this latter case

$$\eta' \geq \frac{T_1 - T_2}{T_1}.$$

It follows, therefore, that if both the forward and reverse Carnot cycles are performed quasistatically, the efficiencies are the same for both

$$\eta = \frac{T_1 - T_2}{T_1} = \eta'.$$

However, it is conceivable that, in a Carnot engine which may be reversed, both cycles are described in a non-static manner. In this case

$$\eta \leq \frac{T_1 - T_2}{T_1} \leq \eta'.$$

Equation (8.1) was written for the reservoirs of a Carnot engine. If it is written for a smaller system, such as the working fluid of a Carnot engine, the difficulty encountered is that the temperature change as heat is added or subtracted may not be neglected. For a general change of state, an integral would be expected. For example, if a system is taken from a state A to a state B, the change of entropy in a typical increment of the process satisfies

$$\delta S \geq \delta'Q / T,$$

where $\delta'Q$ is the heat absorbed by the system at temperature T.

Summing or integrating over all increments

$$S_B - S_A \geq \int_A^B \frac{d'Q}{T}.$$

In particular, if the system is taken through a cyclic process,

$$\int_{cycle} \frac{d'Q}{T} \leq 0.$$

This is known as the **Clausius inequality**. The equality sign holds if the cycle is performed quasistatically.

These results hold for general cycles.

In practice, heat engines generally fall into one of two classes :- *external combustion engines*, of which a good example is the steam engine, and *internal combustion engines*, of which a good example is the familiar petrol engine. In both types, either a gas or a mixture of gases contained in a cylinder is caused to undergo a series of changes which together make up a cycle. This, in turn, causes a piston to make a shaft rotate against some opposing force. Again in both engines, at some time in the cycle, the gas in the cylinder must have its temperature and pressure raised. In the steam engine this is achieved by use of an outside heat source, while, in the petrol engine, it is achieved by a chemical reaction between the fuel and air, which occurs in the cylinder itself. Actually, in the petrol engine, the combustion of petrol and air occurs explosively with the help of an electric spark.

As far as a steam engine is concerned, it may be approximated well by the so-called **Rankine cycle**. In this cycle, all essentially practical complications are eliminated and the cycle is composed of the following six processes:

(a) adiabatic compression of water to the pressure of the boiler where it is to be heated,

(b) isobaric (that is, constant pressure) heating to boiling point of the water,

(c) isobaric, isothermal vaporisation of water into saturated steam,

(d) isobaric superheating of the steam,

(e) adiabatic expansion of steam into wet steam,

(f) isobaric, isothermal condensation of steam into saturated water.

During the processes (b), (c) and (d), an amount of heat Q_1 enters the system from a hot reservoir; but, during the condensation process (f), an amount of heat Q_2 is given by the system to a colder reservoir. The condensation process is necessary to ensure that the system is returned to its initial state - that is, to ensure that the sequence of processes together constitute a complete cycle. It should be noted, however, that, since heat is always given up during the condensation process, Q_2 may never be zero and the input of heat, Q_1 may never be converted completely into work.

The idealised petrol engine performs a cycle known as the **Otto cycle**. Here the behaviour of a petrol engine is approximated by assuming that the working substance is air which is itself assumed to behave like an ideal gas of constant heat capacities, that all processes are quasistatic and that there is no friction. If these assumptions are made, the Otto cycle is composed of the following six processes:

(a)   quasistatic isobaric intake of air at atmospheric pressure, during which the volume varies from zero to V_1 at the temperature of the outside air, θ_1 say.

(b)   quasistatic adiabatic compression of the air, during which the temperature will rise from the temperature of the outside air to a new temperature θ_2 according to the equation

$$\theta_1 V_1^{\gamma-1} = \theta_2 V_2^{\gamma-1}$$

where V_2 is the new volume.

(c)   quasistatic constant volume increase of temperature (to a new temperature θ_3) and pressure, brought about by the absorption of a quantity of heat Q_1 from an external reservoir. It is this process which approximates the effect of the explosion in the actual petrol engine.

(d)   quasistatic adiabatic expansion which involves a drop in temperature from θ_3 to θ_4 and with the volume returning the value V_1 according to

$$\theta_3 V_2^{\gamma-1} = \theta_4 V_1^{\gamma-1}.$$

(e)   quasistatic constant volume decrease in temperature (to the original temperature θ_1) and pressure, brought about by the rejection of heat Q_2 to an external reservoir. This process approximates the return to atmospheric pressure upon opening the exhaust valve.

(f)   quasistatic isobaric exhaust at atmospheric pressure, during which the volume falls from V_1 to zero, with the temperature remaining at the value θ_1.

The two processes, (a) and (f) obviously cancel each other. Two of the remaining four processes involve the flow of heat; these are (c) in which a quantity of heat Q_1 is absorbed , and (e) in which a quantity of heat Q_2 is rejected. Since the working fluid has been assumed to be an ideal gas of constant heat capacities, it follows that

$$Q_1 = \int_{\theta_2}^{\theta_3} C_V \, d\theta = C_V (\theta_3 - \theta_2)$$

and

$$Q_2 = \int_{\theta_4}^{\theta_1} C_V d\theta = C_V \left(\theta_4 - \theta_1 \right),$$

where Q_2 is regarded as a positive number; that is, the expression for Q_2 represents the *positive* amount of heat given up *to* the external reservoir.

Hence, for this case, the efficiency is given by

$$\eta = 1 - \frac{Q_2}{Q_1} = 1 - \frac{\theta_4 - \theta_1}{\theta_3 - \theta_2},$$

where equation (8.2) has been used.

Also, the equations for the two adiabatics may be combined to give

$$\frac{\theta_4 - \theta_1}{\theta_3 - \theta_2} = \left(\frac{V_2}{V_1} \right)^{\gamma - 1}.$$

Therefore, an alternative expression for the efficiency is

$$\eta = 1 - \left(\frac{V_2}{V_1} \right)^{\gamma - 1}.$$

In practice, it is found that the ratio (V_1 / V_2) may not exceed a value of about 10 since, if it is larger, the temperature rise upon compression of the air and petrol mixture is large enough to cause an explosion before the spark: that is, preignition can occur.

In both the examples considered - the steam engine as approximated by the Rankine cycle and the internal combustion engine as approximated by the Otto cycle - simplifying assumptions have been made. All the troublesome effects, such as acceleration, friction and heat loss by conduction, have been eliminated. If taken into account, all would have the effect of lowering the efficiency of the engine. This brief examination of just two actual heat engines has been intended to be merely an introduction to this more practical side of thermodynamics. More details of these, as well as of other practical examples of heat engines, may be found, for example, in *Heat and Thermodynamics* by M.W.Zemansky and also in *Thermal Power Cycles* and *The Conversion of Heat into Work,* both by G.H.A.Cole.

Exercises D

(1) The working fluid of a thermodynamic engine is an ideal gas of constant heat capacities. It works quasistatically in a cyclic process as follows:

(a) isothermal expansion at temperature T_1 from volume V_1 to volume V_2 ;

(b) cooling at constant volume from temperature T_1 to temperature T_2 ;

(c) isothermal compression at temperature T_2 from volume V_2 to volume V_1 ;

(d) heating at constant volume from temperature T_2 to temperature T_1 .

Obtain expressions for the amount of heat, Q_1 , supplied to the gas in steps (a) and (d) , and the amount, Q_2 , rejected in steps (b) and (c).

What does (Q_1 - Q_2) represent physically and why is it reasonable to define the efficiency of the engine as η = (Q_1 - Q_2)/Q_1 ?

For this cycle, show that

$$\eta < (T_1 - T_2)/T_1 .$$

(2) A classical ideal gas of constant heat capacities is used as the working fluid for a quasistatic Carnot cycle as illustrated:

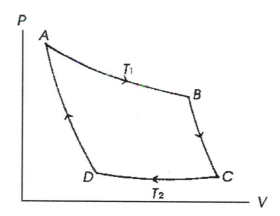

Here *AB* and *CD* are isothermals, *BC* and *DA* are adiabatics. Show that

(a) the work done on the isothermals is

$$AT_1 \log (V_B /V_A) - AT_2 \log (V_C /V_D),$$

(b) $$V_B /V_A = V_C /V_D ,$$

(c) the work done on the adiabatics cancels so that the total work done by the fluid during a cycle is given by

$$W = A (T_1 - T_2) \log (V_B / V_A).$$

(d) Find the efficiency as defined in question (1).

(3)   Two fluids A and B of fixed volumes and constant heat capacities C_1 and C_2 are initially at temperatures T_1 and T_2, with $T_1 > T_2$, respectively. The fluids are adiabatically insulated from one another. A quasistatically acting Carnot engine uses A as a heat source and B as a heat sink and acts between the two fluids until they reach a common temperature T_0.

Obtain expressions for T_0 and for the work done by the Carnot engine.

If a common temperature is established by allowing direct heat flow between A and B, what is the final temperature and what is the change in entropy ?

9

Negative Temperatures and the Second Law

It may appear strange, at first sight, to introduce the seemingly abstruse idea of *negative temperatures* into what is meant to be an essentially undergraduate-level text. However, following the introduction of the definition of absolute temperature in terms of the partial derivative of the entropy with respect to the internal energy in equation (7.5), and noting the reference to the influence of negative temperatures on the equivalence of the Kelvin and Clausius forms of the Second Law, it seems appropriate to examine the concept at this point. It might be noted at the outset that, provided the idea is viewed with an open mind, it is not in fact too abstruse a notion.

If considered from a purely theoretical thermodynamic point of view, it follows, from equation (7.5), that the only requirement for the existence of negative temperatures is that the entropy S should not be restricted to being a monotonically increasing function of the internal energy U; at any point for which the slope of the curve of entropy plotted against internal energy becomes negative, (7.5) shows that the temperature is negative. Normally, in thermodynamics it is not assumed that the entropy increases monotonically with the internal energy since such an assumption is simply not necessary for the development of the subject. However, if no system could be conceived which possessed the property of entropy decreasing with internal energy, there would be little point in pursuing the idea of negative temperature further. Such systems may be both devised theoretically and closely realised experimentally. By using results in statistical mechanics, it may be shown that, for systems of elements in thermal equilibrium which are such that each element of the system has an upper limit to its maximum energy, the plot of entropy against internal energy may have negative values for the slope. This point is well illustrated by considering the case when only two energy states are available to the elements of a system. In this case, when all the elements are in the lowest energy state, the system has its lowest energy and this state is clearly a highly ordered state for the system, and so corresponds to an entropy value of zero. (This argument clearly depends crucially on the link between entropy and the idea of order. A full discussion of this topic is outside the scope of this text but the present brief allusion to it is useful to illustrate how negative temperatures can occur.) Similarly, the largest value of the energy is achieved when all the elements are in the highest energy state. Again, this is a highly ordered state and corresponds also to an entropy value of zero . At intermediate energy values for the system, some elements will be in one energy state and some in the other. Here greater disorder will occur and there will be a correspondingly greater entropy value. Hence, between the lowest and highest energy states of the system, the entropy must achieve a maximum value before decreasing with increasing internal energy. It should be noted that the maximum value of the entropy referred to will correspond to an infinite value for the temperature and so, in cooling from negative to positive temperatures, such a system must pass through an infinite value for the temperature *not* a zero temperature. Also, since negative temperatures occur for higher values of the internal energy than apply for positive values of the temperature, negative temperatures are, in fact, *higher* than positive temperatures. This peculiar result is simply a consequence of equation (7.5).

As mentioned earlier, entropy is not normally assumed to be a monotonically increasing function of the internal energy in the development of thermodynamics simply because such an assumption is not necessary. Hence, the usual theorems and results of thermodynamics apply in both the positive and negative temperature regions provided a little care is taken. The ideas of *work* and *heat* are kept the same for both temperatures but it should be noted that, since

$$d'Q = TdS,$$

a consequence of this is that not all forms of the Second Law can remain unaltered, but more of this in a moment. As indicated earlier, the *hotter* of two bodies will be the one *from* which heat flows when they are brought into thermal contact; the *colder* body will be the one *to* which heat flows. With this definition, any negative temperature is hotter than any positive temperature, while, for two temperatures of the same sign, that with the algebraically larger temperature is the hotter - that is, the temperature scale runs $+0°K, \ldots , +100°K, \ldots , +∞°K, \ldots , -∞°K, \ldots , -100°K, \ldots , -0°K$. Once again, it is reasonable to think this scale seems strange but it must be remembered that it is a consequence of the thermodynamic definition of temperature, - a definition adopted before the idea of negative temperatures was even contemplated. Also, these seemingly odd temperatures appear in very special circumstances which occur only rarely, and the sole reason for mentioning them here is to allow the exact relationship between the Kelvin and Clausius forms of the Second Law to be examined.

The efficiency of a Carnot engine is given by equation (8.2) for both reservoirs at positive temperatures and for both at negative temperatures. It follows from the above discussion, therefore, that , for the negative temperature reservoir case, if heat is absorbed at the hotter temperature,

$$T_2 / T_1 > 1$$

and the efficiency η is seen to be negative. This means that, instead of work being produced when the Carnot engine is run with heat received from the hot reservoir, work *must be supplied* . It follows that, if the cycle is run in the opposite direction, work is produced while heat is transferred from the colder **to** the hotter reservoir. If the heat transferred to the hot reservoir by this reverse cycle is allowed to flow back to the colder reservoir, there exists an engine which, operating in a cycle, produces no effect other than the conversion of a quantity of heat completely into work. This contradicts the Kelvin form of the Second Law but it does *not* contradict the principle of increase of entropy since, from the earlier discussion it is seen that, at negative temperatures, a *decrease* in internal energy corresponds to an *increase* in entropy. It may be noted also that, when a Carnot cycle is operated between two *negative* temperatures so that work is done by the engine while heat is absorbed from the colder reservoir and rejected at the hotter reservoir, the efficiency, as given by equation (8.2), is both positive *and* less than unity. Hence, cyclic heat engines which produce work have efficiencies less than unity irrespective of whether the reservoirs have positive or negative temperatures: that is, they absorb more heat than they produce work.

However, the foregoing discussion shows that, as pointed out by Ramsey {*The Physical Review* **103** (1956) 20-28}, the *Kelvin statement of the Second Law* must be modified in order to take account of negative temperatures. The *modified form* states that

> **in a cyclic process, in the absence of other effects,**
> **heat cannot be converted completely into work**
> **for states of positive absolute temperature and**
> **work cannot be converted completely into heat**
> **for states of negative absolute temperature.**

The Clausius form of the law applies to both positive and negative temperatures and, as mentioned above, so does the so-called entropy formulation - that is, the entropy of an isolated system can never decrease. It remains to consider the connection between this modified form of the Kelvin principle and the Clausius principle.

As was pointed out by Ramsey, it might be felt possible to violate the Clausius principle by constructing an engine which would extract heat from a reservoir and convert it into work without other effects. This work could then be converted into heat which would be transferred to a hotter reservoir. This would indeed violate the Clausius principle. However, from the modified Kelvin principle, it is seen that, at positive absolute temperatures, the first step in this process is impossible and, at negative absolute temperatures, the second step cannot be achieved. Hence, if the modified Kelvin statement of the Second Law does not hold, the Clausius statement could be violated. Therefore, the Clausius statement does imply the modified Kelvin statement.

Now, suppose the Clausius statement is false. In this case, therefore, an engine may be constructed which can transfer heat from one body to another warmer than itself, without producing any other changes. Consider a process in which, at positive absolute temperatures, a quantity of heat is converted partly into work, the remainder having passed from the hot reservoir to a body at a lower temperature. The above Clausius violating engine could be used to return this amount of heat to the hotter reservoir. The combination of these two processes would result in a quantity of heat being converted, in a cyclic process, *completely* into work in violation of the positive temperature part of the modified Kelvin statement of the Second Law. Hence, for positive absolute temperatures, the modified Kelvin principle implies the Clausius principle. Next, consider a process in which, at negative absolute temperatures, a quantity of work is converted *partly* into heat which is delivered to a hot reservoir. In this case, the Clausius violating engine could be used to transfer a further quantity of heat to this hot reservoir from a cooler one - a quantity such that the total amount of heat transferred to the hot reservoir is equal to the total work done. This combination of processes would result in a quantity of work being converted, in a cyclic process, *completely* into heat in violation of the negative temperature part of the modified Kelvin statement of the Second Law. Hence, for negative absolute temperatures also, the modified Kelvin principle implies the Clausius principle. Therefore, it may be concluded that the modified Kelvin principle and the Clausius principle are equivalent.

It should be noted that the entire discussion above indicates that the usual form of the Kelvin principle, which was introduced in Chapter 4, and the Clausius principle cannot be equivalent since the first applies only to positive absolute temperatures and must be modified when negative absolute temperatures are introduced, whereas the second applies equally well to both positive and negative absolute temperatures.

The various statements of the Third Law apply unaltered at negative temperatures when it is realised that the absolute zero of temperature means absolute zero of both positive *and* negative temperature. Thus, the unattainability principle would be that it is impossible in a finite number of steps to reduce any system to the absolute zero of positive temperature ($+0°K$) or to raise any system to the absolute zero of negative temperature ($-0°K$). For completeness, it should be mentioned that both the Zeroth and First Laws apply equally well for both positive and negative absolute temperatures.

10

Phase Transitions

In the familiar phase transitions, such as melting, vaporization and sublimation, as well as in some less familiar transitions, the temperature and pressure remain constant, while the volume and entropy both change. Consider n moles of some material in phase i with molar entropy s^i and molar volume v^i and suppose that ,after the phase transition, the material finds itself in a new phase f with molar entropy s^f and molar volume v^f . (Here by a molar quantity is meant a quantity per mole, where the definition of a mole was given in Chapter 3). Suppose the fraction of the initial phase that has been transformed into the final phase at any moment is x. Then, the entropy and volume of the mixture at any moment are given by

$$S = n(1-x)s^i + nxs^f$$

and

$$V = n(1-x)v^i + nxv^f.$$

If the phase transition takes place reversibly, the heat transferred per mole is given by

$$l = T(s^f - s^i).$$

Here the heat transferred is denoted by l and is usually called a **latent heat,** where a latent heat is the quantity of heat required to change the phase of a substance without an accompanying change of temperature.Latent heat is, of course, that heat which is absorbed by a solid during melting or emitted by a liquid during solidification, and there are latent heats associated with, for example, the liquid - gas transition also.It is molecular energy that, in a melting process, has changed from the kinetic to the potential form and that is why it does not register on a thermometer, which is concerned only with the average kinetic energy of molecules. A familiar example of such a melting process is provided by the melting of ice. The molecules in an ice crystal are more closely bound than they are in the liquid state. These molecules are pulled apart in the melting by work being performed on them; work which increases their potential energy. Alternatively, when a liquid solidifies, this molecular energy changes back from potential to kinetic energy. In the case of water freezing, the molecular potential energy, mentioned above, changes back into kinetic energy and some heat is emitted. The existence of such a latent heat indicates that there is an entropy change.

Also, since

$$dg = -sdT + vdp,$$

where g is the Gibbs function,

$$s = -\left(\frac{\partial g}{\partial t}\right)_p, v = \left(\frac{\partial g}{\partial p}\right)_T,$$

and the familiar phase transitions may be characterised by one of the following equivalent statements:

(a)   there are changes of entropy and volume,

or

(b)   the first-order derivatives of the Gibbs function change discontinuously.

Any phase change satifying these requirements is known as a **phase change of the first order**. It might be noted also that the Gibbs function remains constant during a reversible process taking place at constant temperature and pressure; that is, during a reversible isothermal and isobaric process. Hence, for a change at temperature T and pressure p,

$$g^i = g^f$$

and for a change at temperature $T + dT$ and pressure $p + dp$,

$$g^i + dg^i = g^f + dg^f.$$

Therefore,

$$dg^i = dg^f$$

or

$$-s^i\, dT + v^i\, dp = -s^f\, dT + v^f\, dp.$$

Hence,

$$\frac{dp}{dT} = \frac{s^f - s^i}{v^f - v^i} = \frac{l}{T\left(v^f - v^i\right)}.$$

This equation, the famous **Clapeyron equation**, applies to any first-order change of phase or transition that takes place at constant temperature and pressure.

It is useful to note that, for a phase change such as is envisioned here, the temperature variation of the constant pressure heat capacity, C_p , is as shown in Fig.1:

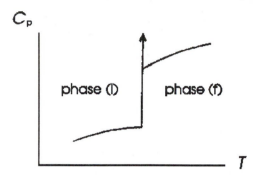

Fig.1

It is seen that the behaviour of the constant pressure heat capacity, C_p , is significant in that, for the *mixture of the two phases during the phase transition*, it is infinite. This is so since both the temperature, T, and the pressure, p, are constant during the transition. Therefore,

$$C_P = T\left(\frac{\partial S}{\partial T}\right)_P \to \infty .$$

since, when p is constant, $dT = 0$. Again, since $dp = 0$ when T is constant, it follows that

$$\alpha_p = \frac{1}{V}\left(\frac{\partial V}{\partial T}\right)_p \quad \text{and} \quad \kappa_T = -\frac{1}{V}\left(\frac{\partial V}{\partial p}\right)_T$$

tend to infinity during the transition also. However, it must be realised that these three statements hold *only* when **both** phases are present. As is seen clearly from the figure, the constant pressure heat capacity of phase i remains finite up to the transition temperature. It does **not anticipate** the onset of the phase transition by starting to increase dramatically before this temperature is reached. This behaviour is true of first order transitions always, but does not occur for **all** types of transition.

In the first order phase changes just discussed, both the entropy and the volume, which are both first order derivatives of the Gibbs function (or Gibbs free energy), undergo finite changes. However, there are phase changes in which the entropy and volume both remain unalterred. In such phase changes, temperature, pressure and Gibbs function remain unchanged also and so, the enthalpy,H, the internal energy, U, and the Helmholtz free energy, F will remain unalterred as well. If, in these transitions, C_p, α_p and κ_T undergo finite changes, then, since

$$C_p = T\left(\frac{\partial S}{\partial T}\right)_p = T\frac{\partial}{\partial T}\left(-\frac{\partial G}{\partial T}\right) = -T\left(\frac{\partial^2 G}{\partial T^2}\right),$$

and

$$\alpha_p = \frac{1}{V}\left(\frac{\partial^2 G}{\partial T\partial p}\right), \qquad \kappa_T = -\frac{1}{V}\left(\frac{\partial^2 G}{\partial p^2}\right),$$

there would be finite changes in the **second order derivatives** of the Gibbs function. Following a method of ordering phase transitions due to Ehrenfest, such transitions are called **second order phase transitions**.

As far as these second order phase transitions are concerned, there are two equations - first derived by Ehrenfest - which express the constancy of s and v in the same way as Clapeyron's equation expresses constancy of the Gibbs function for a first order transition. In this case, once again consider a reversible process which is both isothermal and isobaric. Then, using the same notation as previously, constancy of the entropy is expressed by

$$ds^i = ds^f$$

that is

$$\left(\frac{\partial s^i}{\partial p}\right)_T dp + \left(\frac{\partial s^i}{\partial T}\right)_p dT = \left(\frac{\partial s^f}{\partial p}\right)_T dp + \left(\frac{\partial s^f}{\partial T}\right)_p dT$$

or

$$-v\alpha_p^i\, dp + T^{-1}C_p^i\, dT = -v\alpha_p^f\, dp + T^{-1}C_p^f\, dT.$$

Therefore,

$$\frac{dp}{dT} = \frac{C_p^f - C_p^i}{Tv\left(\alpha_p^f - \alpha_p^i\right)},$$

where $\left(\partial s/\partial p\right)_T = -\left(\partial v/\partial T\right)_p = -v\alpha_p$.

Similarly, constancy of the volume is expressed by

$$dv^i = dv^f$$

and, using the same approach as when dealing with the condition of constancy of the entropy, this leads to the relation

$$\frac{dp}{dT} = \frac{\alpha_p^f - \alpha_p^i}{\kappa_T^f - \kappa_T^i},$$

where κ_T is as defined previously.

These two equations for dp/dT are the **Ehrenfest relations,** which apply to second order phase transitions in the same way that the Clapeyron equation applies to a first order transition.

At one time, there were thought to be many examples of second order transitions, but now it is known that this is not the case. In fact, it may be that there is only *one* such transition: - the change from superconductivity to normal conductivity in zero magnetic field. It has been found experimentally , by making measurements closer and closer to the transition temperature that , in many of the changes thought to be second order, neither C_p nor α_p achieves a finite value at either the beginning or the end of the transition. However, there are transitions of order higher than one and, of these, possibly the most interesting is the so-called **lambda transition** in which

(a)   temperature, pressure and Gibbs function remain constant,

(b)   entropy and volume - and, therefore, internal energy, enthalpy and Helmholtz free energy also, - remain constant,

(c) C_p, α_p, κ_T are infinite.

The two figures 2 and 3 indicate the crucial difference between the second order transition and the lambda transition.

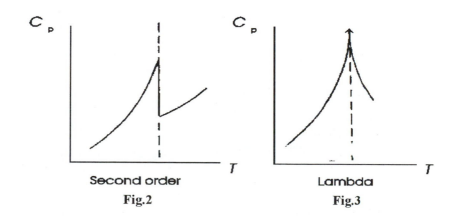

Second order

Fig.2

Lambda

Fig.3

Among the many examples of lambda transitions are:-

(1)   "order - disorder" transformations in alloys,

(2)   the onset of ferroelectricity in certain crystals,

(3)   the transition from ferromagnetism to paramagnetism at the Curie point,

(4)   a change of orientation of an ion in a crystal lattice,

(5)   the transition from ordinary liquid helium (liquid *He* 1) to superfluid helium (liquid *He* 11) at a temperature and corresponding pressure called a *lambda point*.

As was noted earlier, as a substance, in any one phase, approaches the temperature at which a first order transition is to occur, C_p remains finite up to the transition temperature. It becomes infinite only when a small amount of the second phase is present and, before this, its behaviour shows *no evidence* of any premonition of the coming event. However, in the case of a lambda transition, as is seen from figure 3, C_p starts to rise before the transition point is reached as though the substance, in the form of one phase only, *anticipated* the coming phase transition.

Again, as with Chapter 8 which was concerned with thermodynamic cycles, this chapter on phase transitions has been concerned with the very basic theory behind a theoretical description of these physical phenomena. Nowhere has attention been focussed on the vast amount of experimentation carried out in this area; experimentation concerned with the measurement of vapour pressure, with the entire process of vaporization, with the liquifaction of gases, with the onset of transitions other than those of the first order, and much more. Indeed, it is outside the scope of this book to cover such matter but it is important and of interest. Many of these experimental details may be found in *Heat and Thermodynamics* by M.W.Zemansky.

11

Thermodynamic Equilibrium and Stability

In Chapter 5, the basic inequality of thermodynamics

$$d'Q \leq TdS$$

was derived. For an isolated system, $d'Q = 0$ and this inequality becomes

$$dS \geq 0. \tag{11.1}$$

where the equality sign will hold for quasistatic processes, the inequality sign for non-static processes. Since the equilibrium state is, by definition, a state which is stable against spontaneous changes, this inequality indicates that the equilibrium state is a state of maximum entropy. This extremum principle for the entropy may be used to derive several important results relating to the equilibrium state for an isolated system, as well as to the stability of that state.

For simplicity, attention will be resticted to the case where the entropy, S, is a function of the internal energy, U, and the volume, V, only. Consider a closed, composite system consisting of two simple systems separated by a movable wall which permits the passage of heat through it (that is, it is a **diathermal** wall). Suppose the two systems possess internal energies U_1 and U_2 respectively and volumes V_1 and V_2 respectively. Also suppose that these may change only subject to the conditions

$$U_1 + U_2 = \text{const.} \tag{11.2}$$

and

$$V_1 + V_2 = \text{const.} \tag{11.3}$$

For local equilibrium, no change in the entropy is allowed to result from any infinitesimal virtual processes consisting of transfer of heat across the wall or of any movement of the wall itself. Hence

$$dS = 0$$

where

$$dS = \left(\frac{\partial S_1}{\partial U_1}\right)_{V_1} dU_1 + \left(\frac{\partial S_1}{\partial V_1}\right)_{U_1} dV_1 + \left(\frac{\partial S_2}{\partial U_2}\right)_{V_2} dU_2 + \left(\frac{\partial S_2}{\partial V_2}\right)_{U_2} dV_2,$$

S_1 and S_2 being the entropies of the two systems.

From equations (11.2) and (11.3), it follows that

$$dU_1 + dU_2 = 0 \quad \text{and} \quad dV_1 + dV_2 = 0 \tag{11.4}$$

and so

$$dS = \left(\frac{1}{T_1} - \frac{1}{T_2}\right)dU_1 + \left(\frac{p_1}{T_1} - \frac{p_2}{T_2}\right)dV_1 = 0,$$

where it has been noted that

$$\left(\frac{\partial S}{\partial U}\right)_V = \frac{1}{T} \quad \text{and} \quad \left(\frac{\partial S}{\partial V}\right)_U = \frac{p}{T}.$$

Since dU_1 and dV_1 are independent variables, dS vanishing implies

$$T_1 = T_2 \quad \text{and} \quad p_1 = p_2.$$

This result confirms what would have been expected that, in equilibrium, both the temperatures and pressures of the two systems making up the composite system are equal. Equality of the temperatures implies thermal equilibrium, while equality of the pressures implies mechanical equilibrium. However, it should be noted that equality of the pressures follows only **after** equality of the temperatures has been established. Hence, thermal equilibrium must always precede mechanical equilibrium. If initially the entropy had been assumed to depend on the total number of particles in the systems also, a third term would have appeared in the equation for dS -

$$\left(\frac{\mu_2}{T_2} - \frac{\mu_1}{T_1}\right)dN_1,$$

and that would have lead to equality of the chemical potentials of the two systems. This equality would have implied chemical equilibrium but it too would only have been established **after** thermal equilibrium.

It now remains to consider the conditions governing the stability of the equilibrium state discussed above. Again attention will be restricted to the case where the entropy of each system is dependent only on the internal energy and the volume. However, in this case ,consider two identical systems, each with its entropy given by an equation of the form $S = S(U,V)$, which are assumed separated by a movable, diathermal wall once again. If , within the composite system, an amount of energy dU flows from one subsystem to the other, subject to the constraint that the total energy remains constant, and if the wall moves so that the volume of one subsystem increases by an amount dV subject to the total volume remaining constant, then the total entropy change involved is

$$\{S(U + dU, V + dV) - S(U,V)\} + \{S(U - dU, V - dV) - S(U,V)\},$$

where the terms enclosed by { } parentheses are respectively the entropy change for the subsystem gaining energy and increasing in volume, and the entropy change for the second subsystem.

If the equilibrium state is to be a state of maximum entropy, this change must be less than, or equal to, zero; that is,

$$S(U + dU, V + dV) + S(U - dU, V - dV) - 2 S(U,V) \leq 0.$$

Expanding the first two terms in a Taylor series leads to

$$\left(\frac{\partial^2 S}{\partial U^2}\right)(dU)^2 + 2\left(\frac{\partial^2 S}{\partial U \partial V}\right)dUdV + \left(\frac{\partial^2 S}{\partial V^2}\right)(dV)^2 \leq 0.$$

As is shown in the appendix, this condition implies that, for a maximum,

$$\frac{\partial^2 S}{\partial U^2}\frac{\partial^2 S}{\partial V^2} - \left[\frac{\partial^2 S}{\partial U \partial V}\right]^2 \geq 0 \quad \text{and} \quad \frac{\partial^2 S}{\partial U^2} \leq 0.$$

The second of these inequalities may be written

$$\frac{\partial^2 S}{\partial U^2} = \frac{\partial}{\partial U}\left(\frac{1}{T}\right) = -\frac{1}{T^2}\left(\frac{\partial T}{\partial U}\right) = -\frac{1}{T^2 C_V} \leq 0,$$

where the heat capacity involved is that at constant volume since, although not stated specifically, all the partial differentiations with respect to U are carried out at constant volume.

It follows immediately that, for this inequality to be satisfied, it is necessary for the constant volume heat capacity to be *positive*. This is the condition for thermal stability. If an excess of heat is added to a volume of fluid, its temperature will increase relative to the surroundings and so, some heat will flow out again if the heat capacity is positive. However, if the heat capacity was negative, the temperature would decrease, more heat would flow into the volume, and an instability would result.

Also, making use of the result of example 5 in exercises C shows that stability requires

$$C_p \geq C_V \geq 0.$$

It remains to examine the first of the inequalities. To derive a condition comparable with the second inequality, which is seen to reduce to a condition involving the constant volume heat capacity, is a little more complicated but provides an excellent opportunity to use the results involving Jacobians introduced in question 8 of Exercises A.

The first inequality itself may be expressed in terms of a Jacobian as follows:-

$$\frac{\partial^2 S}{\partial U^2}\frac{\partial^2 S}{\partial V^2} - \left(\frac{\partial^2 S}{\partial U \partial V}\right)^2 = \begin{vmatrix} \frac{\partial}{\partial U}\left(\frac{\partial S}{\partial U}\right) & \frac{\partial}{\partial V}\left(\frac{\partial S}{\partial U}\right) \\ \frac{\partial}{\partial U}\left(\frac{\partial S}{\partial V}\right) & \frac{\partial}{\partial V}\left(\frac{\partial S}{\partial V}\right) \end{vmatrix}$$

$$= \partial\left(\frac{\partial S}{\partial U},\frac{\partial S}{\partial V}\right) / \partial(U,V)$$

$$= \partial\left(\frac{1}{T},\frac{p}{T}\right) / \partial(U,V) \geq 0 .$$

By using other properties of Jacobians discussed in the question mentioned, it is seen that the inequality may be transformed easily to

$$\partial\left(\frac{1}{T},\frac{p}{T}\right) / \partial(U,V) = \left[\partial\left(\frac{1}{T},\frac{p}{T}\right) / \partial(T,V)\right]\left[\partial(T,V)/\partial(U,V)\right]$$

$$= -\frac{1}{T^3}\left(\frac{\partial p}{\partial V}\right)_T \frac{1}{C_V} \geq 0 .$$

Since, from above, it is known that the constant volume heat capacity, C_V, is positive, it follows from this latter expression that

$$\left(\frac{\partial p}{\partial V}\right)_T \leq 0,$$

that is, the isothermal compressibility, κ_T, given by

$$\frac{1}{V\kappa_T} = -\left(\frac{\partial p}{\partial V}\right)_T,$$

must be positive.

Hence, it follows that stability requires

$$C_V \geq 0 \quad \text{and} \quad \kappa_T \geq 0,$$

that is, the constant volume heat capacity and the isothermal compressibility are both positive in a stable system. Therefore, addition of heat at constant volume necessarily increases the temperature of a stable system and decreasing the volume isothermally necessarily increases the pressure of a stable system.

As a final point, it might be noted that the physical content of the stability criteria is known as **Le Chatelier's Principle** and, according to this principle, the criterion for stability is that , if an inhomogeneity should develop in a system, it should induce a process which tends to eradicate that inhomogeneity.

12

Concavity of the Entropy and Negative Heat Capacities

In Chapter 11, it was seen that positivity of the heat capacity was one of the conditions ensuring stability of a thermodynamic system. This arose out of the requirement that, at equilibrium, the entropy of a thermodynamic system must have a maximum value. If , for example, a function of one variable is under discussion, it is well-known from ordinary differential calculus that the conditions to be satisfied by a function at a maximum are that its first derivative should be zero and its second derivative should have a negative value. This second condition is interpreted also as indicating that the function is *concave*; - indeed, this interpretation holds whether the first order derivative is zero or not. Similarly, for a function of more than one variable, that function is said to be concave with respect to one of its independent variables if its second order derivative with respect to that variable is negative. If the entropy of a thermodynamic system is taken to be a function of the internal energy and the volume as well as of other variables, then, in the approach to equilibrium, concavity with respect to the internal energy (that is, positive heat capacity) ensures heat flow from high to low absolute temperatures, while concavity with respect to the volume (that is, positive isothermal compressibility) guarantees that the pressure will not increase with the volume. It is interesting to realise that this was Max Planck's criterion for an admissible expression for the entropy as a function of the average energy of an oscillator in his theory of black body radiation published in 1901. Hence, concavity would seem to be an important property of the entropy. Indeed, in an alternative approach to thermodynamics advocated by Callen - *Thermodynamics and an Introduction to Thermostatistics* - concavity of the entropy is taken as the starting point for a derivation of the thermodynamic stability conditions.

In the above discussion, when the second order partial derivative of the entropy with respect to the internal energy is evaluated, all remaining independent variables are held constant. One of these variables would be the total number of particles. Hence, the heat capacity referred to is one evaluated for a fixed total number of particles; that is, it is one evaluated for a *closed* system. It follows that, since systems naturally tend towards equilibrium - that is, towards an entropy maximum - negative heat capacities are not allowed for closed systems. If this was not the case, the above shows that the possibility of violating the Clausius form of the Second Law would exist.

Support for this assertion may be gained by examining results for two systems in thermal contact coming to thermal equilibrium at a common temperature T. Consider two *closed*, isolated systems - one with heat capacity C_1 and at a temperature T_1 , the other with heat capacity C_2 and at a temperature T_2; with $T_2 < T_1$. If these systems are put into thermal contact with one another and achieve thermal equilibrium at temperature T, then, by conservation of energy

$$C_1 (T - T_1) + C_2 (T - T_2) = 0$$

or

$$T = \frac{C_1 T_1 + C_2 T_2}{C_1 + C_2} ,$$

this result holding for both positive and negative heat capacities.

If

$$a = \frac{C_1}{C_1 + C_2} \quad \text{and} \quad b = \frac{C_2}{C_1 + C_2} ,$$

the equation for T may be written

$$T = aT_1 + bT_2 \tag{12.1}$$

where

$$a + b = 1.$$

It remains to consider several cases:

Case 1

If both heat capacities are positive, $a > 0$, $b > 0$ and (12.1) gives

$$\begin{aligned} T &= aT_1 + (1 - a)T_2 \\ &= T_2 + a(T_1 - T_2) \\ &> T_2 \end{aligned}$$

and

$$\begin{aligned} T &= (1 - b)T_1 + bT \\ &= T_1 - b(T_1 - T_2) \\ &< T_1 . \end{aligned}$$

Therefore, in this case,

$$T_1 > T > T_2$$

and

$$C_1 (T - T_1) < 0 , \quad C_2 (T - T_2) > 0.$$

Hence, in this case, it is seen that heat flows from the higher to the lower temperature.

Case 2

If both heat capacities are negative, a $> 0, b > 0$ and, using the same argument as for Case 1, it follows, once again, that

$$T_1 > T > T_2.$$

However, in this case, it is seen also that

$$C_1 (T - T_1) > 0, \quad C_2 (T - T_2) < 0;$$

that is, heat flows from the lower to the higher temperature in violation of the Second Law of Thermodynamics.

Case 3

If one heat capacity is positive and the other negative, then either

$$a > 0, b < 0 \text{ or } a < 0, b > 0.$$

If $a > 0, b < 0$, (12.1) gives

$$T = T_2 + a(T_1 - T_2) > T_2$$

and

$$T = T_1 - b(T_1 - T_2) > T_1$$

so that

$$T > T_1 > T_2.$$

The situation considered in this case may be achieved if

$$C_1 < 0, \quad C_2 > 0, \quad C_1 + C_2 < 0$$

or

$$C_1 > 0, \quad C_2 < 0, \quad C_1 + C_2 > 0.$$

In the first of these

$$C_1 (T - T_1) < 0, \quad C_2 (T - T_2) > 0$$

so heat flows from the higher to the lower temperature.

However, in the second

$$C_1 (T - T_1) > 0, \quad C_2 (T - T_2) > 0$$

so heat flows from the lower to the higher temperature in violation of the Second Law.

Finally, the case $a < 0$, $b > 0$ may be achieved if

$$C_1 > 0, \quad C_2 < 0, \quad C_1 + C_2 < 0$$

or

$$C_1 < 0, \quad C_2 > 0, \quad C_1 + C_2 > 0.$$

A similar argument shows that, in the first of these, heat flows from the higher to the lower temperature but, in the second, it flows from the lower to the higher temperature in violation of the Second Law.

It would seem that the logical conclusion to draw from this discussion is that, when negative heat capacities are involved, the only allowable case is that when the two heat capacities are of different sign and their sum is negative. However, while this may seem eminently reasonable mathematically, is it so reasonable physically? In practice, if a closed system has a negative heat capacity, how can it be ensured that this system comes into thermal contact *only* with other closed systems having positive heat capacities such that the sum of the two heat capacities is negative? Obviously this cannot be ensured and so, the only conclusion that can be drawn from this discussion is that closed systems must have positive heat capacities; - the existence of closed systems with negative heat capacities *could* lead to violations of the Second Law! It is for reasons such as these that the characterising property of the entropy is felt by some to be concavity; that is, concavity embodies the essence of the Second Law.

However, it might be claimed - not unreasonably - that neither of the above arguments, concerning the allowability of negative heat capacities for closed systems, is conclusive, but, if one appeals to more detailed arguments of statistical thermodynamics, which are beyond the scope of this text, further reinforcement for the above view is found. A full discussion of the various points involved, together with a complete list of references, may be found in *Statistical Physics:A Probabilistic Approach*, with further details in *Thermodynamics of Extremes*, both books being by B.H.Lavenda. The only final conclusion that can be reached safely when *all* the evidence and arguments are collected together is that the essence of the Second Law is concavity and so closed systems may not possess negative heat capacities.

This assertion concerning negative heat capacities could cause alarm, at first sight, in some quarters, since the possibility of a star possessing a negative heat capacity has been accepted for many years. However, there is no conflict. The above discussion is concerned with **closed** systems and, since stars are essentially **open** systems, they are not covered by it. As was seen earlier, the constant volume heat capacity may be written

$$C_V = T\left(\frac{\partial S}{\partial T}\right)_V ,$$

which, for a *closed* system, is equal to $\left(\frac{\partial U}{\partial T}\right)_V$. For an *open* system, the Euler relation is seen, from chapter 7, to be

$$TdS = dU + pdV - \mu dN ,$$

and so, for an open system, the expression for the heat capacity becomes

$$C_V = T\left(\frac{\partial S}{\partial T}\right)_V = \left(\frac{\partial U}{\partial T}\right)_V - \mu\left(\frac{\partial N}{\partial T}\right)_V ,$$

where μ and N represent chemical potential and number of particles respectively.

Since the second term on the right-hand side of this equation may be either positive or negative, the sign of the heat capacity in this case remains indeterminate. Hence, the heat capacity of an *open* system *could* be negative. This does not contradict the earlier discussion, nor does it lead to violations of the Second Law . However, an *open* system cannot be isolated and, if such a system and its surroundings are in equilibrium and are considered together as a composite system, that composite system will be a *closed* system possessing a positive total heat capacity.

Often when negative heat capacities are discussed, there is a tacit assumption that systems possessing such heat capacities are allowable and attention is confined to drawing conclusions based on this assumption. However, here it has been demonstrated quite clearly that the existence of *closed* systems with negative heat capacities could lead to violations of the Second Law of Thermodynamics; only *open* systems may have negative heat capacities.

To conclude this chapter, consider once again an ideal classical gas undergoing a polytropic change;- that is, consider the situation introduced in example 10 of Exercises B.

For an ideal classical gas,

$$pV = RT$$

and Joule's Law

$$\left(\frac{\partial U}{\partial V}\right)_T = 0$$

holds.

Also, for such a gas, as was seen in Chapter 3,

$$C_p - C_V = p\left(\frac{\partial U}{\partial V}\right)_p = R.$$

Now consider an ideal classical gas of constant heat capacities C_p , C_V undergoing a quasistatic change for which $d'Q = CdT$, where C is a constant. In this case,

$$d'Q = dU + pdV$$

$$= \left(\frac{\partial U}{\partial T}\right)_V dT + \left\{\left(\frac{\partial U}{\partial V}\right)_T + p\right\}dV$$

that is

$$CdT = C_V dT + pdV$$

$$= C_V dT + \frac{RT}{V}dV$$

or

$$(C_V - C)\frac{dT}{T} = -(C_p - C_V)\frac{dV}{V}.$$

This integrates to give

$$TV^{n-1} = \text{constant,}$$

where $n = (C_p - C)/(C_V - C)$.

With n defined in this way, it follows that

$$C = \frac{(n-\gamma)}{(n-1)}C_V$$

where $\gamma = C_p / C_V$.

Obviously, C may be negative if $1 < n < \gamma$.

Hence, it would appear that an ideal classical gas of constant heat capacities may have a negative heat capacity along so-called polytropic paths described by $TV^{n-1} = $ constant if $1 < n < \gamma$. However, *all* the above results are *independent* of the Second Law and, as has been shown above, introduction of this law results in the exclusion of negative heat capacities for closed systems. Hence, in the above discussion, both C and C_V must be positive and so, either

$$n > 1 \text{ and } n > \gamma$$

or

$$n < 1 \text{ and } n < \gamma .$$

This result is not really surprising since, as Chandrasekhar points out in his book on stellar structure, the above situation is only an ordinary ideal classical gas undergoing a particular type of change.

13

Black Hole Entropy and an Alternative Model for a Black Hole

The discussion in the previous chapter raises serious questions concerning several results in very common use in present day physics; results related to black holes and to string theory. In retrospect, it seems inevitable that the similarity between Hawking's area theorem which asserts that, in any process involving black holes, the total event horizon area cannot decrease, and the well-established thermodynamic result concerning the increase of entropy in isolated systems was one which could not be ignored for long. If a connection was to be established, the question remaining was what function of the area was to be identified with the entropy of a black hole? The simplest choice compatible with Hawking's theorem was to set the black hole entropy proportional to the area of the event horizon itself. This choice was made and the identification is accepted quite widely now, being regarded by some as a contribution to "conventional wisdom".

Since, in the case of a Schwarzschild black hole, the area of the event horizon is proportional to the "irreducible" mass of the black hole, the entropy is assumed given by

$$S = kM^2 \Big/ 2\sigma_m^2 \qquad (13.1)$$

where M is the "irreducible" mass of the black hole and $\sigma_m^2 = ch\big/G$, with c the speed of light, h Planck's constant and G the gravitational constant, is the so-called Planck mass. However, bearing in mind all the thermodynamics that has been discussed so far in this text, is this expression for the black hole entropy correct?

One worrying aspect of equation (13.1) is that the expression is not homogeneous of degree one; or, in other words, the entropy expression is not extensive. This in itself does not bring the validity of the expression into question. However, it does mean that some results used quite frequently in thermodynamic manipulations are available for use no longer. Referring back to Chapter 7, it may be remembered that the so-called Euler relation between the various thermodynamic variables was derived after the entropy had been assumed extensive. Subsequently, the extremely important Gibbs-Duhem relation was deduced. Obviously, neither of these relations would be available for use with the proposed black hole entropy expression given by (13.1).

Again, it may be that the "irreducible" mass M is assumed to adopt the role normally played by the internal energy U or, as is usually the case in black hole thermodynamics, since the original dependence of the black hole entropy is on M, the internal energy is associated with the "irreducible" mass via the relation

$$U = Mc^2 . \tag{13.2}$$

If this identification is used in (13.1), the resulting equation gives

$$\frac{1}{T} = \frac{dS}{dU} = \frac{kU}{c^4 \sigma_m^2}$$

and so, in terms of the temperature, T, (13.1) becomes

$$S = \frac{c^4 \sigma_m^2}{2kT^2} .$$

From this expression, it follows immediately that the heat capacity is given by

$$C = T\frac{dS}{dT} = -\frac{c^4 \sigma_m^2}{kT^2} ;$$

that is, the heat capacity is negative. Although, as mentioned in Chapter 12, negative heat capacities have been accepted in astrophysics for some time, the cases considered there always refer to open systems. Here, although in reality a black hole - if such an object exists - must be an open system, it is treated as a closed system. Bearing in mind the fact that it was shown that, if a closed system had a negative heat capacity, violation of the Second Law could follow, the validity of the proposed entropy expression, (13.1), for a black hole is called into question immediately.

Some might well hold the view that this entire argument calls into question the validity of the Second Law of Thermodynamics, rather than that of the proposed black hole entropy expression. This would not be totally unreasonable since the said law is, in reality, simply a statement of a fact of experience. However, the accepted validity of this law has stood the test of time and , if its validity is to be questioned, it must be done so openly and the fact that it is being challenged must be realised.

Also, is it correct to introduce the internal energy U into the proposed entropy expression (13.1) by using (13.2)? The answer to this question must be "No", since, quite clearly, a so-called rest-energy is being confused with an internal energy.

It should be noted that , considering the case when the entropy depends on the "irreducible" mass M alone has been for simplicity only. The same objections to the

Bekenstein - Hawking entropy expression for a black hole surface in the more general situation where the entropy is a function of the "irreducible" mass M the charge Q and the angular momentum J.

Again, the Hawking area theorem, to which reference has been made already, asserts that, when two black holes coalesce, the area of the resulting event horizon will always be greater than the sum of the areas of the event horizons of the original two black holes; that is, in symbols

$$(M_1 + M_2)^2 > M_1^2 + M_2^2$$

where M_1 and M_2 are the "irreducible" masses of the two original black holes. If this is to be associated with entropy as indicated by (13.1), an obvious problem arises when two systems at the same temperature are brought into thermal contact. In this particular case, the final entropy would be expected to be equal to the sum of the original two entropies. The above shows that, only in the trivial case when one of the masses is zero, will this be so.

Hence, several serious questions may be raised concerning the validity of the commonly accepted expression for the entropy of a black hole. All the objections mentioned here are purely thermodynamic in nature and, at first sight, the possibility of violations of the Second Law is the most damning. Indeed, it would seem that the expression is wrong! However, while it is not unreasonable to criticise the black hole entropy expression (13.1), can an alternative be suggested?

In attempting to suggest an alternative approach, it may be noted that all free degenerate masses up to approximately $4M_0$ (where M_0 is the solar mass) form a hierarchy of objects involving a mechanical equilibrium maintained in each case by the balance between the self- gravity and successively stronger degeneracy forces. It may be noticed that, on a mass-radius plot, all such bodies - including white dwarfs and neutron stars - lie between the limits denoting pure hydrogen and pure iron composition. In this hierarchy, planets and natural satellites, with masses in the range

$$10^{-11} < M/M_0 < 10^{-2},$$

involve non-ionised atomic forces; brown dwarfs, with ionised atoms, increase the upper mass to roughly $M/M_0 < 10^{-1}$; white dwarfs, with electron degeneracy forces, raise it to $M/M_0 < 1.44$, the Chandrasekhar limit; and neutron stars, with neutron degeneracy forces, raise it further to about $M/M_0 < 4$. Often, it is accepted that, beyond this point, there are no sufficiently strong degenerate forces to establish equilibrium with a finite radius. However, there has been speculation that, if a neutron is composed of quarks and if the neutrons in the centre of a neutron star are squeezed further, then these neutrons could decompose into quarks. The quark gas would be degenerate and could support the envelope of the neutron star. Such a model could lead to masses in excess of $4M_0$. Hence, there would seem to be a possibility for extending the mass range for a stable self-gravitating degenerate mass in equilibrium beyond $4M_0$.

Obviously, the form of such an equilibrium may be investigated fully only when the quark dynamics is known, but some preliminary indications of the radius such stars might have may be obtained if the usual analysis of relativistic degenerate objects is relevant, - with the quark mass replacing the electron mass. The relation between the mass and the radius is found to be

$$RM^{\frac{1}{3}} = 1.6 \times 10^{12}.$$

In this way, the relation between mass and radius is defined for objects from the lowest to the highest masses on the basis of a balance between self-gravity and degeneracy forces. More details of the arguments involved in this analysis may be found in *Physics of Planetary Interiors* by G.H.A.Cole.

The full range of quark interactions will lead to an equilibrium distribution of states which allows thermodynamic quantities, including the entropy, to be defined in the usual way. There would seem to be some similarity between such an object and a classical black hole, except that the object would, in this case, have a finite radius which might, however, be very small. It seems conceivable that such a quark star could in fact possess a surface escape speed in excess of the speed of light but it is possible also that such a value for the escape speed could be a limiting value that is never achieved. The special form of the quark coupling certainly suggests that the balance between quark degeneracy and gravitation denies gravity the position of being the ultimate unopposed force for massive bodies.

14

Concluding Remarks

The aim of this book has been to introduce students to the basic principles of thermodynamics. Brief details of the history of the subject have been included at various points in an attempt to draw attention to the background circumstances under which the subject developed and also to recognise the phenomenal contribution made by the founding fathers of thermodynamics; - people such as Rumford, Carnot, Joule, Clausius and Kelvin. Consideration of this background also serves to emphasise the essentially practical, engineering basis of thermodynamics. It is reassuring to have any subject placed on a solid foundation but it is dangerous to forget the origins of a topic, if they are practical, in favour of more abstract theoretical foundations ; - the two should go side by side. If the physics of a situation is forgotten, problems of interpretation of theoretical results can arise quite easily. Here, the approach has been to use a modified form of that introduced by the mathematician Constantin Carathéodory at the instigation of his colleague the physicist Max Born. However, while the approach has been more mathematical than that adopted in many texts, physics has been kept to the fore at all times - the mathematics has been simply a tool. This is a very important point to note. Mathematics is both beautiful and exciting in its own right but it has a vitally important role to play as a tool in a great many areas. In situations where mathematics is primarily a tool, any results derived from the mathematics must be scrutinised carefully to ensure that they possess a *genuine* physical interpretation. This is the basis of the question posed in Chapter 13 concerning the existence of black holes and brings us back to the point raised a little earlier concerning the origins of thermodynamics and its more abstract theoretical foundations. The whole idea of a black hole arose from a singularity occurring in a mathematical model of a physical situation and it was not unreasonable to attempt a "physical" explanation for this singularity. However, until the existence of black holes is proved beyond doubt, either observationally or experimentally, the question of their physical existence must remain: although, in the meanwhile, they can remain godsends for science fiction writers. It might be noted that, although a possible alternative model for a black hole has been mentioned in Chapter 13, it is rated as no more than that - a possibility! The proposed model for a quark star, if valid, could quite easily lead to a situation where the maximum escape speed allowed has to be less than the speed of light. Indeed, in his book *Thermodynamics of Extremes*, Bernard Lavenda speculates that **no** entropy *or* temperature may be attributed to a black hole. He goes on to note that this brings to mind a quotation from Eddington that "there is a magic circle which no measurement can bring us inside". This seems an intriguing point to ponder!

As indicated at the beginning of Chapter 13, another area of thermodynamic controversy is provided by *string theory*. According to modern string theory, the expression for the entropy of a massive string configuration of energy E is

$$S = bE - a\ln E$$

where a and b are constants. It is seen quite easily that this entropy expression leads to a *negative* value for the heat capacity of the system. Hence, if a massive string configuration is to be regarded as a closed system - and there is not even a hint to the contrary in the above entropy expression - the accepted entropy expression would need to be examined carefully since, in its present form, it leads to possible violation of the Second Law of Thermodynamics in just the same way as the widely accepted entropy expression for a black hole does.

As if these two "errors in thermodynamics" are not enough, the relatively modern theory which attempts to resolve some of the fundamental problems arising in the standard big-bang model for the origin of the universe also runs into trouble thermodynamically. The theory released an assumption of adiabaticity and this resulted in the so-called "inflationary" scenario which supposes the supercooling of the universe leads to a period of exponential growth when the latent heat of the phase transition is released. This may be seen to lead to an increase in the entropy of the universe. However, a major difficulty exists. For those familiar with some of the results of the theory of relativity, it may be remembered that the Einstein equations resulting from the Robertson - Walker metric are

$$\ddot{R} = -\frac{4\pi}{3} G(\varepsilon + 3p)R$$

and

$$\left[\frac{\dot{R}}{R}\right]^2 + \frac{k}{R^2} = \frac{8\pi}{3} G\varepsilon$$

where $k = +1$, -1, or 0 depending on whether the universe is closed, open or flat respectively. If the second of the above equations is differentiated with respect to t and the second derivative is eliminated between the resulting equation and the first of the above equations,

$$\frac{d(\varepsilon R^3)}{dt} + p\frac{d(R^3)}{dt} = 0,$$

where ε is the energy density and p the pressure, results. Comparing this equation with the thermodynamic result

$$Td\left(sR^3\right) = d\left(\varepsilon R^3\right) + pd\left(R^3\right)$$

where s is the entropy density and sR^3 the total entropy in a volume whose radius of curvature is R, shows that Einstein's equations imply adiabaticity; that is,

$$d(sR^3) = 0.$$

Hence, *no* criterion for non-adiabatic growth can arise from Einstein's equations.

In this book, the basic laws of thermodynamics have been introduced and discussed, the theory has been extended to cover non-equilibrium situations, heat engines (which, in a sense, provided the impetus for the whole subject to arise) have been discussed as have phase transitions and questions of stability; and finally, the thermodynamics of black holes has been used to introduce the reader to an area which is of immediate widespread interest these days and which is thermodynamically controversial. Although this final topic is controversial, it is, nevertheless, one in which questions need to be asked, and answers given. It is hoped that Chapters 12 and 13 of this present text will help, with the thermodynamic background provided in the earlier chapters, provoke and stimulate discussion of this and, indeed, other areas where basic results of this important, wide-ranging subject are accepted as valid without too much questioning.

Appendix

Here it is intended to develop a brief introduction to the theory of functions of more than one variable. It may be noted from the outset that this type of function is extremely common in the physical sciences: for example, the pressure, p, of a dilute gas is related to its volume, V, and temperature T by $p = RT/V$, where R is a constant. Hence, p is a function of the **two** variables T and V. In general, if a variable quantity z depends on the values of two other variable quantities x and y: $z = f(x,y)$. If x and y may vary independently of one another, they are said to be **independent variables**, while z is said to be the **dependent variable**.

Partial derivatives

If z depends on x and y as indicated above, and x is changed by an increment δx while y is kept constant, then z will change by an increment

$$\delta x = f(x+\delta x, y) - f(x, y).$$

The partial derivative of z with respect to x is defined by

$$\lim_{\delta x \to 0} \frac{\delta z}{\delta x} = \lim_{\delta x \to 0} \frac{f(x+\delta x, y) - f(x, y)}{\delta x}$$

by analogy with the definition of the derivative of a function of one variable. The limit on the right-hand side is the partial derivative of the function $z = f(x,y)$ with respect to x and is denoted by $\partial z / \partial x$ or f_x . Clearly the partial derivative is calculated using the same rules as are employed for ordinary derivatives, provided y is treated as a constant throughout.

Similarly for the partial derivative with respect to y. Obviously the idea may be extended to functions of more than two varibles in which case all the independent variables, other than the one with repect to which the partial derivative is sought, are treated as constants throughout.

Example

Let (x,y) be the rectangular coordinates of a point in a plane, and (r,θ) the polar coordinates so that

$$x = r\cos\theta \ , \ y = r\sin\theta \qquad\qquad (a)$$

Here x and y are given in terms of the independent variables r and θ . Hence,

$$\partial x / \partial r = \cos\theta, \partial y / \partial r = \sin\theta$$

$$\partial x / \partial \theta = -r \sin\theta, \partial y / \partial \theta = r \cos\theta.$$

Again, equations (a) may be solved for (r, θ) in terms of (x, y) to give

$$r = (x^2 + y^2)^{1/2}, \theta = \tan^{-1}(y/x).$$

In these equations the independent variables are x and y and so

$$\partial r / \partial x = x(x^2 + y^2)^{-1/2} = x/r = \cos\theta, \partial r / \partial y = \sin\theta,$$

$$\partial \theta / \partial x = -y/(x^2 + y^2) = -y/r^2 = -(\sin\theta)/r, \partial \theta / \partial y = (\cos\theta)/r.$$

If $z = f(x, y)$, then, in general, the first partial derivatives will be functions of x and y also and may be differentiated partially again with respect to either of the variables. The four second partial derivatives of z are

$$\frac{\partial}{\partial x}\left(\frac{\partial z}{\partial x}\right) = \frac{\partial^2 z}{\partial x^2}, \frac{\partial}{\partial x}\left(\frac{\partial z}{\partial y}\right) = \frac{\partial^2 z}{\partial x \partial y},$$

$$\frac{\partial}{\partial y}\left(\frac{\partial z}{\partial x}\right) = \frac{\partial^2 z}{\partial y \partial x}, \frac{\partial}{\partial y}\left(\frac{\partial z}{\partial y}\right) = \frac{\partial^2 z}{\partial y^2}.$$

Here the order of the symbols $\partial x, \partial y$ from right to left indicates the order of differentiation.

If $z = f(x, y)$, these partial derivatives may be written f_{xx}, f_{xy}, f_{yx}, f_{yy} respectively. In this notation, the order of the suffices from right to left indicates the order of differentiation.

It should be noted that, in general, $f_{xy} \neq f_{yx}$. However, it may be shown that, if f_x and f_y exist in the neighbourhood of a point (a,b) and if f_x and f_y are differentiable at (a,b) then, at the point (a,b), $f_{xy} = f_{yx}$. These conditions are *usually* satisfied in practice, and so, it is even more important to realise that, in general, $f_{xy} \neq f_{yx}$.

The Chain Rule

Suppose $z = f(x, y)$ and let δz be the small change in z corresponding to the small independent changes δx in x and δy in y, then

$$\delta z = f(x+\delta x, y+\delta y) - f(x, y)$$

$$= \frac{\{f(x+\delta x, y+\delta y) - f(x, y+\delta y)\}\delta x}{\delta x} + \frac{\{f(x, y+\delta y) - f(x, y)\}\delta y}{\delta y}.$$

But

$$\lim_{\delta x \to 0} \frac{f(x+\delta x, y+\delta y) - f(x, y+\delta y)}{\delta x} = \frac{\partial}{\partial x} f(x, y+\delta y)$$

and, as $\delta y \to 0$ also, this becomes

$$\frac{\partial}{\partial x} f(x, y) = \frac{\partial z}{\partial x}.$$

Similarly,

$$\lim_{\delta y \to 0} \frac{f(x, y+\delta y) - f(x, y)}{\delta y} = \frac{\partial z}{\partial y}.$$

Thus, in the limit as $\delta x \to 0$ and $\delta y \to 0$,

$$dz = \frac{\partial z}{\partial x} dx + \frac{\partial z}{\partial y} dy \qquad (A.1)$$

gives the differential of z. This definition follows in the same way as for a function of one variable.

Now suppose $z = f(x,y)$ but x and y, instead of being independent variables, are both functions of a single variable t. Corresponding to a given increment δt in t, there will be increments δx and δy in x and y and an increment δz in z. In this case, it is seen that

$$\frac{dz}{dt} = \frac{\partial z}{\partial x} \frac{dx}{dt} + \frac{\partial z}{\partial y} \frac{dy}{dt}. \qquad (A.2)$$

The latter two results may be generalized further:- If z is given as a function of the variables $x_1, x_2, ..., x_n$, then

$$dz = \sum_{i=1}^{n} \frac{\partial z}{\partial x_i} dx_i,$$

and if $x_1, x_2, ..., x_n$ are each given functions of $t_1, t_2, ...,$ then

$$\frac{\partial z}{\partial t_i} = \sum_{j=1}^{n} \frac{\partial z}{\partial x_j} \frac{\partial x_j}{\partial t_i},$$

for $i = 1,2,......$

These results are referred to as the **chain rule**.

Example

If $z = f(x,y)$ where $2x = e^u + e^v$ and $2y = e^u - e^v$, show that

$$\frac{\partial z}{\partial u} + \frac{\partial z}{\partial v} = x\frac{\partial z}{\partial x} + y\frac{\partial z}{\partial y}.$$

Now

$$\frac{\partial z}{\partial u} = \frac{\partial z}{\partial x}\frac{\partial x}{\partial u} + \frac{\partial z}{\partial y}\frac{\partial y}{\partial u} = \frac{1}{2}e^u\left(\frac{\partial z}{\partial x} + \frac{\partial z}{\partial y}\right).$$

Similarly,

$$\frac{\partial z}{\partial v} = \frac{1}{2}e^v\left(\frac{\partial z}{\partial x} - \frac{\partial z}{\partial y}\right).$$

Therefore,

$$\frac{\partial z}{\partial u} + \frac{\partial z}{\partial v} = \frac{1}{2}\frac{\partial z}{\partial x}\left(e^u + e^v\right) + \frac{1}{2}\frac{\partial z}{\partial y}\left(e^u - e^v\right) = x\frac{\partial z}{\partial x} + y\frac{\partial z}{\partial y}.$$

It might be noted that, if z is a function of x and y where y itself is a function of x, z becomes a function of x alone when y is expressed in terms of x. This case is covered by (A.2) with $t = x$, and so

$$\frac{dz}{dx} = \frac{\partial z}{\partial x} + \frac{\partial z}{\partial y}\frac{dy}{dx}.$$

Here dz/dx is the total derivative of z with respect to x; that is, the derivative with respect to x of the function obtained by substituting for y in z its expression as a function of x. $\partial z / \partial x$ is the partial derivative of z with respect to x when y is kept constant.

When y is defined as a function of x by an equation $f(x,y) = 0$, y is said to be an **implicit function** of x. In such a case, dy/dx may be found as follows:-

By definition, y is a function of x such that, when substituted in $f(x,y)$, the resulting identity is

$$f\{x,y(x)\} = 0.$$

Since f is identically zero, its total derivative is zero, and so

$$\frac{df}{dx} = \frac{\partial f}{\partial x} + \frac{\partial f}{\partial y}\frac{dy}{dx} = 0$$

or

$$f_x + f_y \frac{dy}{dx} = 0,$$

that is

$$dy/dx = -f_x /f_y .$$

Example

Consider the function

$$f(x, y) = \frac{x^2}{a^2} + \frac{y^2}{b^2} - 1 = 0.$$

$$f_x = 2x/a^2 , \quad f_y = 2y/b^2$$

and so

$$dy/dx = -xb^2 /ya^2 .$$

Homogeneous Functions

A function $f(x_1 ,x_2 ,....,x_m)$ is said to be *homogeneous* of degree n in the variables $x_1 ,x_2 ,....,x_m$ if

$$f(x_1 t,x_2 t ,....,x_m t) = t^n f(x_1 ,x_2 ,....,x_m).$$

For example, $(x^2 + y^2 + z^2)$ is homogeneous of degree 2, and $(x + y)/(x^4 + z^4)$ is homogeneous of degree -3.

Appendix

Theorem

If $f(x_1, x_2, ..., x_m)$ is a homogeneous function of degree n in $x_1, x_2, ..., x_m$, then

$$x_1 \frac{\partial f}{\partial x_1} + x_2 \frac{\partial f}{\partial x_2} + ... + x_m \frac{\partial f}{\partial x_m} = nf.$$

(This is a theorem originally due to Euler.)

Proof

Let $x_1 = \alpha_1 t, x_2 = \alpha_2 t, ..., x_m = \alpha_m t$ so that

$$u = f(x_1, x_2, ..., x_m) = f(\alpha_1 t, \alpha_2 t, ..., \alpha_m t) = t^n f(\alpha_1, \alpha_2, ..., \alpha_m).$$

The function u is a function of the variable t and, by differentiating the two equivalent forms $f(\alpha_1 t, \alpha_2 t, ..., \alpha_m t)$ and $t^n f(\alpha_1, \alpha_2, ..., \alpha_m)$, two forms of the derivative du/dt are obtained.

$$\frac{du}{dt} = \frac{\partial u}{\partial x_1} \frac{dx_1}{dt} + \frac{\partial u}{\partial x_2} \frac{dx_2}{dt} + ... + \frac{\partial u}{\partial x_m} \frac{dx_m}{dt}$$

$$= \alpha_1 \frac{\partial u}{\partial x_1} + \alpha_2 \frac{\partial u}{\partial x_2} + ... + \alpha_m \frac{\partial u}{\partial x_m}.$$

Also

$$du / dt = nt^{n-1} f(\alpha_1, \alpha_2, ..., \alpha_m).$$

On multiplying each expression by t it is seen that

$$\alpha_1 t \frac{\partial u}{\partial x_1} + \alpha_2 t \frac{\partial u}{\partial x_2} + ... + \alpha_m t \frac{\partial u}{\partial x_m} = nt^n f(\alpha_1, \alpha_2, ..., \alpha_m)$$

$$= nf(\alpha_1 t, \alpha_2 t, ..., \alpha_m t).$$

Hence the required result.

Taylor's Theorem for a Function of Several Variables

It is assumed that this theorem has been met already for a function of one variable; that is, for a function $f(t)$, the Taylor series expansion about the point $t = 0$ is

$$f(t) = f(0) + \frac{df}{dt}\bigg|_{t=0} t + \frac{1}{2!}\frac{d^2 f}{dt^2}\bigg|_{t=0} t^2 + \ldots = \sum_n \frac{1}{n!} f^n(0) t^n$$

where $f^n(0)$ is the nth. derivative of $f(t)$ evaluated at $t=0$.

The Taylor series expansion for a function of several variables may be deduced from the above as follows: Suppose the function is $u(x)$ and that the variables x are given as functions of the single variable t by

$$x = x_0 + th \tag{A.3}$$

where x_0 and h are fixed. Then $u(x)$ becomes a function of the single variable t, so that

$$u(t) = u(0) + \frac{du}{dt}\bigg|_{t=0} t + \frac{1}{2!}\frac{d^2 u}{dt^2}\bigg|_{t=0} t^2 + \ldots \tag{A.4}$$

If each component of x is denoted by x_i, then from (A.3)

$$x_i = (x_0)_I + th_i$$

and

$$dx_i / dt = h_i .$$

Using the chain rule gives

$$\frac{du}{dt} = \sum_i \frac{\partial u}{\partial x_i} \frac{dx_i}{dt} = \sum_i \frac{\partial u}{\partial x_i} h_i$$

and

$$\frac{d^2 u}{dt^2} = \sum_i \frac{d}{dt}\left(\frac{\partial u}{\partial x_i} h_i\right) = \sum_{i,j} \frac{\partial}{\partial x_j}\left(\frac{\partial u}{\partial x_i} h_i\right) h_j = \sum_{i,j} \frac{\partial^2 u}{\partial x_j \partial x_i} h_i h_j$$

together with expressions for the higher order derivatives.

In (A.4), these differentials are evaluated at $t=0$; that is, when $x = x_0$. Substituting into (A.4) and putting $t = 1$ yields

$$u(x_0 + h) \;=\; u(x_0) + \sum_i \left.\frac{\partial u}{\partial x_i}\right|_{x=x_0} h_i + \frac{1}{2!}\sum_{i,j} \left.\frac{\partial^2 u}{\partial x_i \partial x_j}\right|_{x=x_0} h_i h_j + \dots$$

This is the Taylor series expansion for a function of n variables x about the point x_0. The expansion for the function $u(x)$ as a power series in x is obtained by putting $x_0 = 0$ and $h = x$ in the last equation. Hence,

$$u(x) \;=\; u(0) + \sum_i \left.\frac{\partial u}{\partial x_i}\right|_{x=0} x_i + \frac{1}{2!}\sum_{i,j} \left.\frac{\partial^2 u}{\partial x_i \partial x_j}\right|_{x=0} x_i x_j + \dots$$

Extreme Values of Functions of Several Variables

Once again, it is assumed that the theory of extreme values for functions of one variable is understood, and that it remains to consider the extension to the case of functions of several variables. For simplicity, attention will be restricted to a function of two independent variables, but it should be noted that the theory is applicable regardless of the number of independent variables.

Consider the function $f(x,y)$, which is a function of the two independent variables x and y. Such a function is said to possess a maximum value at the point (a,b) if

$$f(a + h, b + k) - f(a,b) < 0$$

for all sufficiently small positive or negative values of h and k. A minimum is defined with the inequality reversed.

Applying Taylor's theorem gives

$$f(a+h,b+k) = f(a,b) + \left(h\frac{\partial f}{\partial x} + k\frac{\partial f}{\partial y} \right) + \frac{1}{2}\left(h^2\frac{\partial^2 f}{\partial x^2} + 2hk\frac{\partial^2 f}{\partial x \partial y} + k^2\frac{\partial^2 f}{\partial y^2} \right) + \dots$$

where all the partial derivatives are evaluated at $x = a$, $y = b$. For convenience, introduce the following notation for these partial derivatives

$$\frac{\partial f}{\partial x} = f_a, \frac{\partial f}{\partial y} = f_b, \frac{\partial^2 f}{\partial x^2} = f_{aa}, \frac{\partial^2 f}{\partial x \partial y} = f_{ab}, \frac{\partial^2 f}{\partial y^2} = f_{bb},$$

then

$$f(a + h, b + k) \;=\; f(a,b) + (hf_a + kf_b) + (h^2 f_{aa} + 2hk f_{ab} + k^2 f_{bb})/2 + \dots \quad (A.5)$$

Hence, for sufficiently small values of h and k, the sign of $[f(a + h, b + k) - f(a,b)]$ is the sign of $(hf_a + kf_b)$. However, for a maximum (or minimum) value of $f(x,y)$ at (a,b), the sign of this term must be negative (or positive) for all independent positive or negative values of h and k. If $k = 0$, changing the sign of h would change the sign of $(hf_a + kf_b)$ unless $f_a = 0$. Similarly, if $h = 0$, f_b would need to equal zero. Thus, a necessary condition for $f(x,y)$ to possess an extreme value - a maximum or a minimum - at the point (a,b) is

$$f_a = 0 \text{ and } f_b = 0;$$

that is, both f_x and f_y should be zero at the point (a,b).

If these conditions are satisfied

$$f(a + h, b + k) - f(a,b) = (h^2 f_{aa} + 2hk f_{ab} + k^2 f_{bb})/2 + \ldots.$$

and again for sufficiently small h and k, the sign of the right-hand side depends on the sign of

$$(h^2 f_{aa} + 2hk f_{ab} + k^2 f_{bb}).$$

Now put $h = \rho \cos\theta, k = \rho \sin\theta$, where ρ is positive. The signs of h and k depend now on the value of θ. Also

$$h^2 f_{aa} + 2hk f_{ab} + k^2 f_{bb} = \rho^2(f_{aa} \cos^2\theta + 2f_{ab} \cos\theta \sin\theta + f_{bb} \sin^2\theta). \quad (A.6)$$

Suppose that not all of f_{aa}, f_{ab}, f_{bb} are zero since, if they were, the above term would vanish and it would be necessary to consider the next term in the Taylor series. If $f_{aa} = f_{bb} = 0$ but $f_{ab} \neq 0$, the sign of (A.6) is determined by the sign of $f_{ab} \cos\theta \sin\theta$, and this changes sign if θ (and therefore k) changes sign. Under these conditions, the function may have neither a maximum nor a minimum value at the point (a,b). In fact, $f(x,y)$ is said to have a **saddle point** at (a,b).

If $f_{aa} \neq 0$, the expression on the right-hand side of (A.6) may be written

$$\rho^2\{(f_{aa} \cos\theta + f_{ab} \sin\theta)^2 + (f_{aa}f_{bb} - f_{ab}^2)^2 \sin^2\theta\}/f_{aa} \quad (A.7)$$

or, if $f_{bb} \neq 0$, it may be written

$$\rho^2\{(f_{bb} \sin\theta + f_{ab} \cos\theta)^2 + (f_{aa}f_{bb} - f_{ab}^2) \cos^2\theta\}/f_{bb.} \quad (A.8)$$

Three cases now need to be considered:

(a) $f_{aa}f_{bb} - f_{ab}^2 > 0.$

If this condition is satisfied

$$f_{aa}f_{bb} > f_{ab}^2 > 0$$

and so, f_{aa} and f_{bb} are of the same sign. Also, the signs of the above two expressions (A.7) and (A.8) are seen to be the same as the sign of f_{aa} (or f_{bb}). Hence, there is a maximum value if $f_{aa} < 0$ and a minimum value if $f_{aa} > 0$.

(b) $f_{aa} f_{bb} - f_{ab}^2 < 0.$

If this condition is satisfied, the expression (A.7) is *not* of invariable sign. When $\theta = 0$ it is positive but, when $\theta = \tan^{-1} (-f_{aa} / f_{ab})$, it is negative. Hence, once more the function has a stationary value which is neither a maximum nor a minimum; that is, a saddle point. It might be noted that the situation $f_{aa} = f_{bb} = 0, f_{ab} \neq 0$ is covered by this case.

(c) $f_{aa} f_{bb} - f_{ab}^2 = 0.$

In this case, the expression (A.7) has the same sign as f_{aa} except possibly for

$$\theta = \tan^{-1} (-f_{aa} / f_{ab}).$$

Further investigation is necessary to determine the nature of the stationary value under these conditions. In general, this is not easy and the case will not be considered further here.

Answers and Solutions to Exercises

Exercises A

(1) $4t^3$.

(2)   (a) -1, (b) $(y^2 x^2 + y^2 + 2xy^2 - 3x^3 + 4x^3 y)/x^2 y^2$.

(3)   (a) $\dfrac{\partial u}{\partial x} = \dfrac{1}{y}\sec^2\!\left(\dfrac{x}{y}\right), \dfrac{\partial u}{\partial y} = -\dfrac{x}{y^2}\sec^2\!\left(\dfrac{x}{y}\right).$

(b) $\partial f / \partial r = 2r\sin^2\theta + 3r^2 ,\;\; \partial f / \partial\theta = 2r^2 \sin\theta\cos\theta$.

(c) $\partial u / \partial r = 3r^2 + t - 1, \partial u / \partial s = 2st, \partial u / \partial t = s^2 + r - 3.$

(d) $\partial f / \partial p = 2p\log q\,\exp(p^2 \log q), \partial f / \partial q = p^2 q^{-1}\exp(p^2 \log q).$

(4)   The equation may be written

$$V = At/p + B$$

and so

$$\left(\frac{\partial V}{\partial p}\right)_t = -\frac{At}{p^2} = \frac{(B-V)}{p}, \left(\frac{\partial V}{\partial t}\right)_p = \frac{A}{p} + \frac{dB}{dt} .$$

Also

$$dV = \left(\frac{A}{p} + \frac{dB}{dt}\right)dt - \frac{At}{p^2}\,dp.$$

(5)   Suppose z is a function of the independent variables x and y; then

$$dz = \left(\frac{\partial z}{\partial x}\right)_y dx + \left(\frac{\partial z}{\partial y}\right)_x dy \qquad \text{(a)}$$

If y is considered to be a function of x and z,

$$dy = \left(\frac{\partial y}{\partial x}\right)_z dx + \left(\frac{\partial y}{\partial z}\right)_x dz .$$

Substituting for dy in (a) gives

$$dz = \left(\frac{\partial z}{\partial x}\right)_y dx + \left(\frac{\partial z}{\partial y}\right)_x \left\{ \left(\frac{\partial y}{\partial x}\right)_z dx + \left(\frac{\partial y}{\partial z}\right)_x dz \right\} .$$

Comparing coefficients of dz gives

$$1 = \left(\frac{\partial z}{\partial y}\right)_x \left(\frac{\partial y}{\partial z}\right)_x \qquad \text{(b)}$$

and of dx gives

$$0 = \left(\frac{\partial z}{\partial x}\right)_y + \left(\frac{\partial z}{\partial y}\right)_x \left(\frac{\partial y}{\partial x}\right)_z .$$

Rearranging and using (b) gives

$$\left(\frac{\partial x}{\partial y}\right)_z \left(\frac{\partial y}{\partial z}\right)_x \left(\frac{\partial z}{\partial x}\right)_y = -1.$$

(6) $V = ar^b h^{3-b}$

$$\left(\frac{\partial V}{\partial r}\right)_h = bar^{b-1}h^{3-b} ; \left(\frac{\partial V}{\partial h}\right)_r = (3-b)ar^b h^{2-b}$$

$$dV = \left(\frac{\partial V}{\partial r}\right)_h dr + \left(\frac{\partial V}{\partial h}\right)_r dh$$

$$= ar^{b-1} h^{2-b} \{bhdr + (3-b)rdh\}$$

$$= L(r,h)dr + M(r,h)dh.$$

$$\left(\frac{\partial L}{\partial h}\right)_r = ab(3-b)r^{b-1}h^{2-b} = \left(\frac{\partial M}{\partial r}\right)_h.$$

Therefore, dV is exact.

(7)   The solution to this question is a straightforward manipulation.

(8)
$$\left(\frac{\partial E}{\partial T}\right)_\mu = \frac{\partial(E,\mu)}{\partial(T,\mu)} = \frac{\partial(E,\mu)}{\partial(T,N)}\frac{\partial(T,N)}{\partial(T,\mu)} = \frac{\partial(E,\mu)}{\partial(T,N)}\left(\frac{\partial N}{\partial\mu}\right)_T$$

$$= \left\{\left(\frac{\partial E}{\partial T}\right)_N \left(\frac{\partial\mu}{\partial N}\right)_T - \left(\frac{\partial E}{\partial N}\right)_T \left(\frac{\partial\mu}{\partial T}\right)_N\right\}\left(\frac{\partial N}{\partial\mu}\right)_T$$

$$= \left(\frac{\partial E}{\partial T}\right)_N - \left(\frac{\partial E}{\partial N}\right)_T \left(\frac{\partial\mu}{\partial T}\right)_N \left(\frac{\partial N}{\partial\mu}\right)_T$$

$$= \left(\frac{\partial E}{\partial T}\right)_N + \left(\frac{\partial E}{\partial N}\right)_T \left(\frac{\partial N}{\partial T}\right)_\mu,$$

where the result of question (5) has been used.

(9)   If $z = f(x^n y)$ then

$$\partial z/\partial x = f'nx^{n-1}y \quad\text{and}\quad \partial z/\partial y = f'x^n$$

where f' is the derivative of f with respect to its argument.

The required result follows immediately.

Exercises B

(1)   The stationary points of the equation are given by

$$\left(\partial p / \partial V\right)_t = 0,$$ (a)

that is, by

$$p - a/V^2 + 2ab/V^3 = 0.$$ (b)

The stationary value of *this* curve is given by

$$dp/dV = 0,$$

that is, by

$$V = V_{cr} = 3b.$$

From (a) and (b) it follows that

$$p_{cr} = a/27b^2 \; ; \quad t_{cr} = 8a/27Ab.$$

Also, it is seen that, when $V = V_{cr}$, $d^2 p/dV^2$ is negative for the curve represented by (b) and so, the stationary value is a maximum.

From the expressions for V_{cr} , p_{cr} and t_{cr} , it is seen that

$$b = At_{cr} /8p_{cr} \; , \quad a = 3p_{cr} \; (V_{cr})^2 , \quad V_{cr} = 3At_{cr} /8p_{cr} \; .$$

Substituting these in equation (1) leads to the required result.

(2)   For a van der Waals' gas,

$$p = At/(V - b) - a/V^2 = At/V(1 - b/V) - a/V^2$$

that is

$$pV = At + (b - a/At)AtV^{-1} .$$

However, if a and b are *small*, the initial equation shows that

$$V \approx At/p$$

and so

$$pV \approx At + (b - a/At)p + \ldots.$$

Therefore, the *approximate* second virial coefficient is

$$B = b - a/At.$$

(3)   For a van der Waals' gas, the expansion in question (2) shows that

$$\left[\frac{\partial(pV)}{\partial p} \right]_{p=0} = b - \frac{a}{At}$$

which is zero provided $t = t_B = a/Ab$.

Therefore,

$$\frac{t_B}{t_{cr}} = \frac{a}{Ab} \frac{27Ab}{8a} = 3.375$$

(4) $$d(\log V) = \frac{dV}{V} = \frac{1}{V} \left\{ \left(\frac{\partial V}{\partial t} \right)_p dt + \left(\frac{\partial V}{\partial p} \right)_t dp \right\}$$

Comparing this with the given equation gives

$$\alpha_p = \frac{1}{V} \left(\frac{\partial V}{\partial t} \right)_p \quad ; \quad \kappa_t = -\frac{1}{V} \left(\frac{\partial V}{\partial p} \right)_t$$

α_p is the coefficient of volume expansion at constant pressure.

κ_T is the isothermal compressibility and is, in fact, the reciprocal of the isothermal bulk modulus of elasticity.

From its definition, $d(\log V)$ is the exact differential of $\log V$ and so, it follows immediately, using the result given in question (6) of exercises A, that

$$\left(\frac{\partial \alpha_p}{\partial p}\right)_t = -\left(\frac{\partial \kappa_T}{\partial t}\right)_p .$$

(5)   From the given equations,

$$(C_p - C_V)dt = l_V\, dV - l_p\, dp$$

and so

$$d'Q = \frac{C_p l_V}{\left(C_p - C_V\right)}dV + \left\{l_p - \frac{l_p C_p}{\left(C_p - C_V\right)}\right\}dp$$

and so

$$m_V = C_p l_V /(C_p - C_V) \quad ; \quad m_p = -C_V l_p /(C_p - C_V).$$

The second required result follows immediately.

(6) $\Gamma = \dfrac{\alpha_p V}{\kappa_T C_V} = -\dfrac{1}{V}\left(\dfrac{\partial V}{\partial t}\right)_p V.V\left(\dfrac{\partial p}{\partial V}\right)_t C_V = \dfrac{V}{C_V}\left(\dfrac{\partial p}{\partial t}\right)_V = V/\left(\dfrac{\partial U}{\partial p}\right)_V$

since $C_V = \left(\dfrac{\partial U}{\partial t}\right)_V$ and the result of question (5) of exercises A has been used.

Again, using the First Law,

$$d'Q = dU + pdV$$

$$= \left\{ \left(\frac{\partial U}{\partial V} \right)_p + p \right\} dV + \left(\frac{\partial U}{\partial p} \right)_V dp$$

that is

$$m_p = \left(\frac{\partial U}{\partial p} \right)_V .$$

Hence required result.

(7) For a quasistatic adiabatic change

$$pV^\gamma = \text{const.} \quad \text{where} \quad \gamma = C_p / C_V .$$

For a Boyle's Law fluid

$$pV = \theta .$$

Therefore, for a Boyle's Law fluid undergoing quasistatic adiabatic changes at constant γ ,

$$\theta V^{\gamma - 1} = \text{const.} ; \quad \theta^\gamma p^{1-\gamma} = \text{const.}$$

(8) It is seen immediately that

$$dH = dU + pdV + Vdp$$
$$= d'Q + Vdp$$
$$= C_p \, dt + (l_p + V)dp .$$

Also,

$$\left(\frac{\partial H}{\partial p} \right)_t = \left(\frac{\partial U}{\partial p} \right)_t + p \left(\frac{\partial V}{\partial p} \right)_t + V .$$

Therefore,

$$\left(\frac{\partial p}{\partial V}\right)_t \left\{\left(\frac{\partial H}{\partial p}\right)_t - V\right\} - p = \left(\frac{\partial p}{\partial V}\right)_t \left\{\left(\frac{\partial U}{\partial p}\right)_t + p\left(\frac{\partial V}{\partial p}\right)_t\right\} = \left(\frac{\partial U}{\partial V}\right)_t.$$

Again,

$$-\left(\frac{\partial p}{\partial V}\right)_t \left(\mu C_p + V\right) - p = -\left(\frac{\partial p}{\partial V}\right)_t \left\{\left(\frac{\partial t}{\partial p}\right)_H \left(\frac{\partial H}{\partial t}\right)_p + V\right\} - p$$

$$= \left(\frac{\partial p}{\partial V}\right)_t \left\{\left(\frac{\partial H}{\partial p}\right)_t - V\right\} - p = \left(\frac{\partial U}{\partial V}\right)_t.$$

(9) $$\frac{p\mu C_p}{V} = -\frac{p}{V}\left(\frac{\partial H}{\partial p}\right)_t = -\frac{p}{V}\left\{\left(\frac{\partial U}{\partial p}\right)_t + V + p\left(\frac{\partial V}{\partial p}\right)_t\right\}$$

If $pV = At$,

$$\left(\frac{\partial p}{\partial V}\right)_t = -\frac{p}{V}.$$

Hence, in this case,

$$\frac{p\mu C_p}{V} = \left(\frac{\partial p}{\partial V}\right)_t \left(\frac{\partial U}{\partial p}\right)_t = \left(\frac{\partial U}{\partial V}\right)_t.$$

(10)   In this case,

$$d'Q = dU + p\,dV$$

$$= \left(\frac{\partial U}{\partial t}\right)_V dt + \left(\frac{\partial U}{\partial V}\right)_t dV + p\,dV$$

$$= C_V\,dt + p\,dV.$$

Also, for a polytropic change,

$$pV^n = \text{const.}$$

that is

$$AtV^{n-1} = \text{const.}$$

Differentiating gives

$$dV = -\frac{Vdt}{(n-1)t},$$

and so,

$$d'Q = C_V \, dt - \frac{pV}{(n-1)t} dt$$

$$= \frac{\left[(n-1)C_V - A\right]}{(n-1)} dt$$

$$= \left[A - (n-1)C_V\right]\frac{t}{V} dV.$$

Exercises C

(1)   For quasistatic changes,

$$d'Q = TdS = dU + pdV$$

that is,

$$dU = TdS - pdV$$

Then,

$$dF = dU - TdS - SdT$$
$$= -pdV - SdT$$

$$dH = dU + pdV + Vdp$$
$$= TdS + Vdp$$

$$dG = dU + pdV + Vdp - TdS - SdT$$
$$= Vdp - SdT.$$

(2)   *U, F, H* and *G* are all real-valued, differentiable functions and so, the expressions in (1) for *dU, dF, dH* and *dG* are exact differentials. Hence, using the condition for exactness from question (6) of exercises A gives

$$\left(\frac{\partial T}{\partial V}\right)_S = -\left(\frac{\partial p}{\partial S}\right)_V \quad ; \quad \left(\frac{\partial p}{\partial T}\right)_V = \left(\frac{\partial S}{\partial V}\right)_T \quad ; \quad \left(\frac{\partial T}{\partial p}\right)_S = \left(\frac{\partial V}{\partial S}\right)_p \quad ;$$

$$\left(\frac{\partial V}{\partial T}\right)_p = -\left(\frac{\partial S}{\partial p}\right)_T .$$

(3)   For a paramagnetic system kept at constant volume and pressure,

$$d'Q = TdS = dU - BdM.$$

And the analogues of the Maxwell equations are seen to be

$$\left(\frac{\partial T}{\partial M}\right)_S = \left(\frac{\partial B}{\partial S}\right)_M \quad ; \quad \left(\frac{\partial B}{\partial T}\right)_M = -\left(\frac{\partial S}{\partial M}\right)_T \quad ; \quad \left(\frac{\partial T}{\partial B}\right)_S = -\left(\frac{\partial M}{\partial S}\right)_B \quad ;$$

$$\left(\frac{\partial M}{\partial T}\right)_B = \left(\frac{\partial S}{\partial B}\right)_T .$$

Again,

$$TdS = d'Q = \left(\frac{\partial Q}{\partial T}\right)_M dT + \left(\frac{\partial Q}{\partial M}\right)_T dM$$

$$= C_M \, dT + l_M \, dM.$$

$$C_M = \left(\frac{\partial Q}{\partial T}\right)_M = T\left(\frac{\partial S}{\partial T}\right)_M .$$

Similarly,

$$C_B = T\left(\frac{\partial S}{\partial T}\right)_B .$$

Also,

$$(C_B - C_M)dT = l_M \, dM - l_B \, dB$$

where

$$l_M = T\left(\frac{\partial S}{\partial M}\right)_T = -T\left(\frac{\partial B}{\partial T}\right)_M$$

and

$$l_B = T\left(\frac{\partial S}{\partial B}\right)_T = T\left(\frac{\partial M}{\partial T}\right)_B,$$

two of the above Maxwell relations having been used.

Therefore,

$$(C_B - C_M) = l_M \left(\frac{\partial M}{\partial T}\right)_B = -T\left(\frac{\partial B}{\partial T}\right)_M \left(\frac{\partial M}{\partial T}\right)_B$$

$$= T\left[\left(\frac{\partial M}{\partial T}\right)_B\right]^2 \left(\frac{\partial B}{\partial M}\right)_T .$$

(4)
$$TdS = dU + pdV$$

Taking V and T as independent variables, this equation may be written

$$T\left\{\left(\frac{\partial S}{\partial V}\right)_T dV + \left(\frac{\partial S}{\partial T}\right)_V dT\right\} = \left\{\left(\frac{\partial U}{\partial V}\right)_T + p\right\} dV + \left(\frac{\partial U}{\partial T}\right)_V dT$$

and comparing the coefficients of dV gives

$$\left(\frac{\partial U}{\partial V}\right)_T = T\left(\frac{\partial S}{\partial V}\right)_T - p = T\left(\frac{\partial p}{\partial T}\right)_V - p.$$

Once again a Maxwell relation has been used. Now, Joule's Law states that

$$\left(\frac{\partial U}{\partial V}\right)_T = 0 \, ,$$

and so

$$T\left(\frac{\partial p}{\partial T}\right)_V - p = 0 \, ,$$

that is

$$pA \;=\; T$$

where A is a function of V only, and so the equation may be written

$$pg(V) \;=\; T. \tag{a}$$

Suppose the equation of state

$$pV \;=\; f(T) \tag{b}$$

holds also, then (a) and (b) lead to

$$g(V) / V \;=\; T / f(T).$$

Here the left- hand side is a function of volume only while the right-hand side depends on temperature alone. Hence, each side must equal a constant, say A. Then

$$g(V) \;=\; AV \quad \text{and} \quad f(T) \;=\; T / A.$$

(5)   With the usual notation,

$$(C_p - C_V)dT \;=\; l_V \, dV - l_p \, dp$$

and so,

$$C_p - C_V = l_V \left(\frac{\partial V}{\partial T} \right)_p = -l_p \left(\frac{\partial p}{\partial T} \right)_V .$$

But

$$l_V = T \left(\frac{\partial S}{\partial V} \right)_T = T \left(\frac{\partial p}{\partial T} \right)_V$$

$$l_p = T \left(\frac{\partial S}{\partial p} \right)_T = -T \left(\frac{\partial V}{\partial T} \right)_p$$

where Maxwell relations have been used.

Therefore,

$$C_p - C_V = T \left(\frac{\partial p}{\partial T} \right)_V \left(\frac{\partial V}{\partial T} \right)_p = -T \left[\left(\frac{\partial V}{\partial T} \right)_p \right]^2 \left(\frac{\partial p}{\partial V} \right)_T$$

$$= \frac{TV\alpha_p^2}{\kappa_T} ,$$

where the definitions for α_p and κ_T ,introduced earlier, have been used

(6)   As seen in question (4),

$$\left(\frac{\partial U}{\partial V} \right)_T + p = T \left(\frac{\partial S}{\partial V} \right)_T = T \left(\frac{\partial p}{\partial T} \right)_V ,$$

but, for an ideal classical gas,

$$pV = AT \implies \left(\frac{\partial p}{\partial T} \right)_V = \frac{A}{V}$$

and so

$$\left(\frac{\partial U}{\partial V} \right)_T = T \frac{A}{V} - p = 0 ;$$

that is, the internal energy is independent of volume in this case. It follows that

$$dU = \left(\frac{\partial U}{\partial T}\right)_V dT + \left(\frac{\partial U}{\partial V}\right)_T dV = C_V dT$$

Integrating gives

$$U(T) \ - \ U(0) \ = \ C_V \ T.$$

Finally, for the system described, since the weight is placed on the system *gently*, the increase in the internal energy is

$$-\int_{V_1}^{V_2} p_2 dV = -p_2 \int_{V_1}^{V_2} dV = p_2\left(V_2 - V_1\right).$$

However,

$$U(T_2) \ - \ U(T_1) \ = \ C_V \ (T_2 - T_1)$$

and so

$$C_V \ (T_2 \ - \ T_1) \ = \ p_2 \ (V_1 - V_2) \ = \ p_2\left\{\frac{AT_1}{p_1} - \frac{AT_2}{p_2}\right\}$$

$$= \ A\lambda T_1 - AT_2 .$$

Rearranging leads to

$$\frac{T_2}{T_1} = \frac{C_V + A\lambda}{C_V + A} .$$

Exercises D

(1)   For the system under consideration, $pV = AT$, Joule's Law holds and the heat capacities are constant. Hence, the First Law takes the form

$$d'Q = dU + pdV = C_v\, dT + pdV.$$

Therefore, the heat supplied in step (a) is

$$\int pdV = AT_1 \int_{V_1}^{V_2} \frac{dV}{V} = AT_1 \log\!\left(\frac{V_2}{V_1}\right)$$

and, similarly, the heat rejected in step (c) is

$$AT_2 \log\!\left(\frac{V_2}{V_1}\right).$$

Also, the heat supplied in step (d) is

$$\int_{T_2}^{T_1} C_V dT = C_V\left(T_1 - T_2\right)$$

and that rejected in step (b) is, similarly, $C_V(T_1 - T_2)$.
Therefore,

$$Q_1 = AT_1 \log\!\left(\frac{V_2}{V_1}\right) + C_V\left(T_1 - T_2\right)$$

and

$$Q_2 = AT_2 \log\!\left(\frac{V_2}{V_1}\right) + C_V\left(T_1 - T_2\right).$$

However, by the First Law, the working fluid performs an amount of work W on its surroundings which is given by

$$Q_1 = W + Q_2$$

so that

$$W = Q_1 - Q_2.$$

Hence, the given definition of the efficiency, η, is the ratio of the work done to the heat supplied; - which seems reasonable. For the given cycle

$$\eta = \frac{Q_1 - Q_2}{Q_1} = \frac{A(T_1 - T_2)\log\left(\frac{V_2}{V_1}\right)}{AT_1\log\left(\frac{V_2}{V_1}\right) + C_V(T_1 - T_2)} < \frac{T_1 - T_2}{T_1}.$$

(2)   For the system under consideration, $pV = AT$, Joule's Law holds and, on the adiabatics,

$$pV^\gamma = \text{const.}$$

The work done on the isothermal at temperature T_1 is

$$AT_1 \int_{V_A}^{V_B} \frac{dV}{V} = AT_1 \log\left(\frac{V_B}{V_A}\right)$$

and that done on the isothermal at temperature T_2 is

$$-AT_2 \log\left(\frac{V_C}{V_D}\right).$$

For a quasistatic adiabatic process,

$$TV^{\gamma-1} = \text{const.}$$

And, if this is applied to the adiabatics BC and DA , it is seen that

$$\frac{V_A}{V_B} = \frac{V_D}{V_C}.$$

Hence, the work done on the isothermals is $A(T_1 - T_2)\log\left(\dfrac{V_B}{V_A}\right)$.

For an adiabatic, $d'Q = 0$ and so the work done on the adiabatics is $-\int dU$.
Since C_V is constant, it follows that the work done on the adiabatic BC is $C_V(T_1 - T_2)$ and that on the adiabatic DA is $-C_V(T_1-T_2)$, so that the work done on the adiabatics cancels.

Therefore, the total work done during the cycle is given by

$$W = A(T_1 - T_2)\log\left(\frac{V_B}{V_A}\right).$$

However, Joule's Law holds and so, as the gas expands isothermally, its internal energy remains constant; the work done equals the heat obtained to compensate it.

Therefore, for AB,

$$Q_1 = AT_1\log\left(\frac{V_B}{V_A}\right).$$

Then, the efficiency is given by

$$\eta = \frac{W}{Q_1} = \frac{T_1 - T_2}{T_1}.$$

(3)   Since $C_1 = T\left(\dfrac{\partial S}{\partial T}\right)_V$, the entropy lost by A is

$$S_1 = -C_1 \int_{T_1}^{T_0} \frac{dT}{T} = C_1 \log\left(\frac{T_1}{T_0}\right)$$

and that gained by B is

$$S_2 = C_2 \log\left(\frac{T_o}{T_2}\right).$$

If the working substance of the Carnot engine is in the same state finally as it was initially, the entropy gain of the whole system is

$$S_2 - S_1 = \log\left\{\left(\frac{T_0}{T_2}\right)^{C_2}\left(\frac{T_0}{T_1}\right)^{C_1}\right\} = 0.$$

and so

$$T_0 = T_1^a T_2^b$$

where $a = \dfrac{C_1}{C_1 + C_2}, b = \dfrac{C_2}{C_1 + C_2}.$

The internal energy lost by A is in the form of heat:

$$Q_1 = -C_1 \int_{T_1}^{T_0} dT = C_1(T_1 - T_0)$$

and that gained by B is

$$Q_2 = C_2(T_0 - T_2).$$

Therefore, the overall loss of internal energy is

$$Q_1 - Q_2 = C_1 T_1 + C_2 T_2 - (C_1 + C_2)T_0$$

and, by conservation of energy, this must equal the total amount of work done by the Carnot engine.

In the absence of the performance of work, energy conservation yields

$$C_1 (T_1 - T_0) = C_2 (T_0 - T_2)$$

so that

$$T_0 = aT_1 + bT_2$$

with a and b as before.

In this case also, the gain of entropy for the whole system is

$$S_2 - S_1 = (C_1 + C_2) \log \left\{ \left(\frac{T_0}{T_2} \right)^b \left(\frac{T_0}{T_1} \right)^a \right\} = (C_1 + C_2) \log \frac{T_0}{T_1^a T_2^b}$$

since $a + b = 1$.

Therefore,

$$S_2 - S_1 = (C_1 + C_2) \log \left(\frac{aT_1 + bT_2}{T_1^a T_2^b} \right).$$

Glossary

Adiabatic change - one in which there is no thermal interaction.

Black hole - a region in which matter has collapsed to such an extent that even light can escape from it no longer.

Chemical equilibrium - this means there are no chemical reactions within the system.

Closed system - a system whose mass does not alter.

Deformation coordinate - a coordinate (such as volume) whose change implies an alteration or deformation in the size and/or shape of the system.

Extensive variable - one which depends on the size, or extent, of the system; for example, volume, internal energy and number of particles.

First Law - energy is conserved when heat is taken into account.

Heat capacity - a quantity which tells by how much the temperature of a system rises for a given input of heat. (Two heat capacities are encountered commonly; - the constant volume heat capacity C_v and the constant pressure heat capacity C_p)

Inflationary scenario - the theory of the very early universe in which the entire universe, as presently observed, inflated from a minute speck of matter.

Intensive variable - one which does not depend on the size of the system and so, is not extensive.

Internal energy - the energy which depends only on the internal state (as determined by the density and temperature) of a substance.

Isentropic change - a change for which the entropy remains constant. (Note use of the Greek word *isos* meaning *same*)

Isothermal change - a change for which the temperature remains constant.

Latent heat - the quantity of heat necessary to change a substance from one state to another without a change in temperature. (For example, the latent heat of vaporisation is associated with the change from the liquid to the vapour state, while the latent heat of fusion is associated with the change from the solid to the liquid state.)

Mechanical equilibrium - this means there are no unbalanced forces acting on any part of the system or on the system as a whole.

Neutron star - very small type of dead star, only tens of kilometres in diameter, in which the matter is in the form of neutrons crushed together until they touch.

Non-static process - one which is not quasistatic. (see quasistatic change below)

Open system - a system whose mass may alter.

Pressure - force per unit area.

Quasistatic process - an idealised process during which the system passes only through equilibrium states; that is, it consists exclusively of a sequence of thermodynamic equilibrium states.

Second Law -

Clausius form states that it is impossible for heat to be transferred by a cyclic process from a body to one warmer than itself without producing other changes at the same time.

Kelvin form states that it is impossible to transform an amount of heat completely into work in a cyclic process in the absence of other effects.

Temperature - that quantity which tells how hot a system is. (An accurate definition of temperature is possible only on the basis of the Second Law of Thermodynamics.)

Thermal equilibrium - this means there are no temperature differences between parts of the system or between the system and its surroundings.

Thermodynamic equilibrium - to be in thermodynamic equilibrium, a system must satisfy the requirements of thermal, mechanical and chemical equilibrium.

Third Law - the entropy of every system at the absolute zero of temperature may be taken equal to zero.

(Another popular form is that it is impossible to cool any substance to the absolute zero of temperature.)

Volume - the amount of space occupied by a system.

White dwarf - small type of dying star, only a few thousand kilometres in diameter, with a surface temperature in the region of $10^5 \, ^0C$, in the process of cooling off to a cold, dead star.

Zeroth Law - if two systems are separately in thermal equilibrium with a third, then they must be in thermal equilibrium with one another.

List of symbols

C_p constant pressure heat capacity

C_v constant volume heat capacity

F Helmholtz free energy

G Gibbs free energy

H enthalpy

ℓ_p latent heat of pressure increase

ℓ_V latent heat of volume increase

N number of particles

p pressure

Q heat

S entropy

t empirical temperature

T absolute temperature

U internal energy

V volume

W work

α_p coefficient of volume expansion at constant pressure

γ ratio of the constant pressure and constant volume heat capacities

η efficiency

κ_t isothermal compressibility

μ chemical potential

References and suggestions for further reading

Brush, S.G.   *The Kind of Motion we call Heat*
     (North-Holland, 1976)

Born, M.   *Natural Philosophy of Cause and Chance*
     (Dover, 1964)

Callen, H.B.   *Thermodynamics and an Introduction to Thermostatistics*
     2nd. Edition.   (John Wiley, 1985)

Cole, G.H.A.   *Physics of Planetary Interiors*
     (Adam Hilger,1984)
     Thermal Power Cycles
     (Arnold, 1991)
     Engineering Thermodynamics: Gas and Steam Cycles for Converting Heat into Work
     (Albion Publishing, 1996)

Landsberg, P.T.   *Thermodynamics with Quantum Statistical Illustrations*
     (Interscience, 1961)

Lavenda, B.H.   *Statistical Physics:A Probabilistic Approach*
     (John Wiley,1991)
     Thermodynamics of Extremes
     (Albion Publishing, 1995)

Planck, M.   *Treatise on Thermodynamics*

                     (Dover,1945)

Pippard, A.B.   *Classical Thermodynamics*

                     (Cambridge U.P., 1964)

Sears, F.W.   *Thermodynamics*

                     (Addison-Wesley, 1966)

Zemansky, M.W. *Heat and Thermodynamics"*

                     5th. Edition. (McGraw-Hill, 1968)

For the interested reader, as far as the more controversial topics discussed in Chapters 12, 13 and 14 are concerned, more details of the arguments involved may be found in the two books by B. H. Lavenda included in the above list, as well as in the following articles:

B. H. Lavenda and J. Dunning-Davies, *Found. Phys. Lett.*,**3**. 435, (1990).

J. Dunning-Davies, *Found. Phys. Lett.*, **6**, 289, (1993).

B. H. Lavenda and J. Dunning-Davies, *Class. & Quantum Gravity*, **5**, L149, (1988), *Int. J. Theor. Phys.* **29**, 509, (1990).

B. H. Lavenda, *Z. Natur.*, A**45**, 879, (1990).

B. H. Lavenda and J. Dunning-Davies, *Nature*, **368**, 284, (1994).

J. Dunning-Davies, *Trends in Stat. Phys.*, **1**, 23, (1994).

Index